Socio-Legal Aspects of the 3D Printing Revolution

Angela Daly

Socio-Legal Aspects of the 3D Printing Revolution

Angela Daly
Swinburne University of Technology
Hawthorn, Australia

ISBN 978-1-137-51555-1   ISBN 978-1-137-51556-8   (eBook)
DOI 10.1057/978-1-137-51556-8

Library of Congress Control Number: 2016940835

Printed on acid-free paper

This Palgrave Pivot imprint is published by Springer Nature
The registered company is Macmillan Publishers Ltd. London

ACKNOWLEDGEMENTS

I would like to thank Swinburne University of Technology for its support of my research published in this book through a Faculty of Life and Social Sciences Research Development Grant in 2013 which enabled me to start working in the area of 3D printing. Specifically, I would like to thank my director, Professor Julian Thomas, who has steadfastly supported my work in this area, and colleagues Dr Amanda Scardamaglia and Dr Ramon Lobato, who have commented on drafts of this work. Special thanks to Darcy Allen, who was my research assistant on the initial part of this project. Thanks as well to Robbie Fordyce for his comments.

I have presented parts of this research-in-progress in various places in Europe and Australia. In particular, I would like to thank the INFOSOC Working Group at the European University Institute, the NEXA Center for Internet and Society at the Politecnico di Torino and the Digital Media Research Centre at Queensland University of Technology for hosting my seminars and providing useful audience feedback. I would also like to acknowledge the Tilburg Institute for Law, Technology and Society which has hosted me as a research associate during my European trips in order to carry out 3D printing-related research.

CONTENTS

CHAPTER 1

Regulating Revolution: An Introduction to 3D Printing and the Law

Abstract Additive manufacturing or 'three-dimensional (3D) printing' has emerged into the mainstream in the last few years, with much hype about its revolutionary potential as the latest 'disruptive technology' to destroy existing business models, empower individuals, and evade any kind of government control. This book examines the trajectory of 3D printing in practice and how it interacts with various areas of law, including intellectual property (IP), product liability, gun laws, data privacy, and fundamental/constitutional rights. Before the detailed examination of law and 3D printing, this opening chapter introduces 3D printing as a technology, along with some of the high-level themes which permeate its interaction with areas of law. A particular comparison is made with the Internet as this has been, legally speaking, another 'disruptive technology' and also one on which 3D printing is partially dependent.

This chapter will present an overview of the book and its contents, along with where it fits into the current scholarship. The methodology used in the book will be described here, that is, a 'law in society' approach, which has been chosen over a 'black letter law' approach, given the challenges that 3D printing as a decentralised cross-jurisdictional technology and its relationship with the Internet and personal computing pose for the effective enforcement of legal rules.

© The Editor(s) (if applicable) and The Author(s) 2016

A. Daly, *Socio-Legal Aspects of the 3D Printing Revolution*,
DOI 10.1057/978-1-137-51556-8_1

1

The choice of 3D printing as an object of study will be explained: 3D printing's entry into the mainstream and emergence as a possible consumer product as well as its proliferating applications in various areas of life along with the beginnings of academic narratives around the technology from disciplines beyond engineering and manufacturing make it a timely object of study. However, hitherto there are no book-length accounts from a legal perspective on the topic despite the challenges 3D printing presents to certain areas of law and its effective enforcement—a gap this book aims to fill.

A short description will be given of 3D printing itself, and the different yet related technologies comprised in this umbrella term. Initial 3D printing developments in the 1980s will be discussed in brief, along with the significance of patents on restricting the mass availability of 3D printing, the creation of the RepRap movement out of a British university in accordance with free and open-source principles, and, finally, the commodification of 3D printing and its push to consumers.

An outline of the following chapters of the book will then be set out, which explore the legal implications of 3D printing in more depth, examining in particular the issues 3D printing poses for IP law (including its relationship with competition), arms control, product liability, privacy and data protection, as well as the enforceability of all of these regimes.

This Book's Approach

This book looks at 3D printing and its interaction with various areas of law, from a socio-legal perspective. This 'law in context' approach[1] is preferred over a the traditional 'black letter' method due to the dimension that a consideration of how 3D printing is being used in reality adds to the research, whereas a black letter approach would limit consideration merely just to how 'law in the books' applies to the emerging technology. Thus, a political economy approach to the law is taken by this book,[2] in order to uncover how institutions are encountering 3D printing, the extent to

---

[1] W. Twining (2008) 'Law in Context movement' in P. Cane and J. Conaghan (eds.) *The New Oxford Companion to Law* (Oxford: Oxford University Press).

[2] F. Pasquale (2014) 'Symbiotic Law & Social Science: The Case for Political Economy in the Legal Academy, and Legal Scholarship in Political Economy' (Jotwell 5th Anniversary Conference, Miami).

which it is 'freeing' for individuals, and whether (and which) laws can and do govern it.

A transnational approach is taken to the question of jurisdiction: US law is evidently prominent in this book's analysis given it is the locus of much 3D printing activity, where 3D printing companies are headquartered, and whose laws govern the Terms and Conditions of Use of many online 3D printing services such as Thingiverse, the prominent 3D design file-sharing platform.

However, American laws are not the only ones examined here, despite their pre-eminence. As with the Internet, 3D printing is a transnational phenomenon in part due to its leveraging of the Internet—especially when it comes to 3D printing design file distribution and access, which causes problems (albeit not completely insurmountable) for the effective enforcement of one jurisdiction's laws.[3] Accordingly, the European Union (EU) law and its manifestation in one jurisdiction, the UK, are also considered as a site of comparison and contrast, especially on IP, fundamental rights, and data protection. Australian law, another similar common law system, is brought in at times to provide an additional jurisdiction's perspective on 3D printing issues for law, especially in the discussion on gun control, as specific legislative reform has been considered there to address the emergence of 3D printed weapons.

This book does not aim to provide a comprehensive comparison of how different jurisdictions will encounter 3D printing in their legal, economic, and social systems. Indeed, it may well be that 3D printing is actually of more value to people in the Global South, in terms of making a meaningful contribution to their standard of living, than to these (over) developed jurisdictions where mass manufacturing of goods in places such as China still represents a cheaper and more viable option than domestic production.[4] There may also be differing approaches to the enforcement of law, especially in jurisdictions where law enforcement is inadequately

---

[3] J. Reidenberg (2005) 'Technology and Internet Jurisdiction', *University of Pennsylvania Law Review*, 153(6), 1951–1974; D. Svantesson (2014) 'Sovereignty in international law – how the internet (maybe) changed everything, but not for long' *Masaryk University Journal of Law and Technology*, 8(1), 137–155.

[4] See: T. Birtchnell and W. Hoyle (2014) *3D Printing for Development in the Global South* (Basingstoke and New York: Palgrave Macmillan).

resourced.[5] Yet even between the USA and EU, there are important divergences in the law as it applies to 3D printing, which can be seen in particular in Chap. 3's discussion on gun control.

Another important aspect of this book is that its focus is on the domestic- or consumer-oriented usage of 3D printing rather than its industrial-scale applications. The 'liberatory' aspect of 3D printing as a social phenomenon is more likely to be realised in this small-scale use, inasmuch as techniques which previously were too complicated to be performed by individuals or small organisations due to a lack of expertise, resources, and so on can be carried out through the use of 3D printers. Indeed, as will be seen in the discussion throughout this book, this decentralised production using 3D printers poses theoretical if not also practical challenges to various areas of law, which are based on production taking place at a centralised and industrialised level, as well as these products reaching consumers along centralised processes of distribution.

Yet, before these themes are explored in more depth, more explanation of 3D printing as a technology and its origins are warranted.

A Brief History of 3D Printing

What is now popularly known as 3D printing is actually a bundle of technological developments that originally were termed 'rapid prototyping' or 'additive manufacturing' (and these terms are still used, particularly in more technical literature). Essentially, these techniques all permit the relatively cheap and quick creation of a prototype for industrial product development (hence 'rapid prototyping'), and also involve the construction of objects via the building up of material, usually in a layer-upon-layer fashion—hence 'additive' manufacturing as compared to the traditional (retroactively named) 'subtractive manufacturing' by which a piece of raw material is cut into a final shape and size by a process whereby excess and unwanted material is removed.

Conceptual precursors to 3D printers can be found in science fiction, especially the *Star Trek* Replicator (and such a comparison has been

[5] See: R. White (2012) 'Police Cooperation' in M. E. Beare (ed.), *Encyclopedia of Transnational Crime and Justice* (Thousand Oaks: Sage).

encouraged in the literature around 3D printers).[6] However, real-life 3D printing can trace its practical origins to the 1970s. In 1977, an inventor named Wyn Kelly Swainson was granted a patent in the USA for a process whereby a laser is used to solidify liquid plastic to form solid plastic along the path of the beam.[7] This process envisaged the solidification of this plastic in layers in order to produce a 3D object, controlled by a computer. This can be seen as 'the beginning of practical additive manufacture of three-dimensional parts under computer control'.[8] The first patent for stereolithography apparatus (SLA) was issued to Charles (Chuck) Hall in 1983, who went on to co-found 3D Systems which used this technology to introduce the first commercial rapid prototyping system in 1987. In 1987, Carl Deckard from the University of Texas filed a patent in the USA for selective laser sintering (SLS), which was issued in 1989. Finally, Scott Crump, who co-founded Stratasys, filed a patent for fused deposition modelling (FDM), a process currently used by many low-cost consumer-oriented 3D printers.[9]

Various other 3D printing techniques were developed during the 1990s and early 2000s, but all were aimed at industrial applications. Towards the end of the 2000s, 3D printers began to come down in price, with the notable launch of a machine under US$10,000 from 3D Systems in 2007. However, it was around this time that the open-source/open-hardware self-replicating RepRap was launched and began to gain visibility. From 2009 onwards, consumer-level 3D printers were created and put on the market by a number of manufacturers, and by 2012, 3D printing has broken into the mainstream, at least in developed Western markets.

[6] M. Michael (2014) 'Process and Plasticity: Printing, Prototyping and the Prospects of Plastic' in J. Gabrys, G. Hawkins and M. Michael (eds.) *Accumulation: The Material Politics of Plastic* (London: Routledge).

[7] B. Cumptson, M. Lipson, S. R. Marder, J. W. Perry (1999) 'Two-photon or higher-order absorbing optical materials' US Patent Application PCT/US19991008383, http://www.google.com/patents/WO1999053242A1?cl=en, date accessed 10 September 2015.

[8] A. Bowyer (2006) Keynote Address on the RepRap Project (Seventh National Conference on Rapid Design, Prototyping & Manufacturing, High Wycombe), http://reprap.org/wiki/PhilosophyPage, accessed 10 September 2015.

[9] 3D Printing Industry (2014) History of 3D Printing, http://3dprintingindustry.com/3d-printing-basics-free-beginners-guide/history/, accessed 10 September 2015.

The RepRap is an important development in the 3D printing story as it can be seen as an initial attempt to 'democratise' the technology. The project was created by Adrian Bowyer, a Senior Lecturer in mechanical engineering at the University of Bath, and was an initiative to develop a 3D printer that could re-print most of its own components. Around the time of its launch, commercial 3D printers were too expensive for the average consumer in the developed world to afford, with the RepRap providing a much cheaper option, with material costs estimated at €350, such that it was 'accessible to small communities in the developing world as well as individuals in the developed world'.[10] The project releases all of the designs it produces under the GNU General Public licence, in an attempt to transfer free software/Creative Commons principles to hardware—and spurring the 'Open Hardware' movement. Designers are free to modify RepRap designs so long as they share their creations back with the RepRap community.

However, the RepRap also, perhaps unwittingly, spawned commercial offerings of low-cost consumer-oriented 3D printers, most notably those developed by MakerBot. Makerbot's founders met at the NYC Resistor Hackerspace, and 'threw out the self-replication requirement' of RepRap.[11] By 2011, MakerBot had sold several thousand printers; the following year, it attracted significant venture capital funding, and finally was bought by Stratasys.

At the time of writing, the 3D printing industry is maturing, with a significant amount of consolidation around Stratasys (as their purchase of MakerBot demonstrates) and 3D Systems, which are increasingly viewed as the 'Big Two' duopoly in the industry.

An important development alongside 3D printing is 3D scanning, whereby a 3D object can be scanned in order to collect data on its shape and other properties such as appearance. This data can then be used to create 3D digital models (which themselves can be printed on a 3D printer). Similarly to 3D printing, various technologies can be used to undertake this 3D scanning. This scanning can be carried out: by lasers; by touching a probe to various points on the surface of an object; white light scanning; CT scanning; and photo-image-based systems. These techniques have been used industrially and in research for the last few decades, but

---

[10] RepRap (2014) 'About', http://reprap.org/wiki/About, accessed 10 September 2015.

[11] R. Courtland (2013) 'Resources Profile: Bre Pettis' *IEEE Spectrum*, http://spectrum.ieee.org/geek-life/profiles/bre-pettis, accessed: 10 September 2015.

the advent of 3D printing and a need to provide input for these printers have resulted in 3D printer manufacturers such as MakerBot releasing a consumer-oriented (relatively) low-cost scanner called the Digitizer, which takes a scan of an object from multiple angles in order to build a 3D digitised model of said object. At the time of writing, there are various applications in development which turn a smartphone into a 3D scanner,[12] and accordingly lower the cost and increase the accessibility of this technique.

Furthermore, incumbent industries have embraced 3D printing, seemingly to a much greater extent than consumer take-up on the technology, and in sharp contrast to how certain incumbents, particularly from the media and cultural industries, resisted the digitisation and sharing of content from the 1990s.

Among the numerous companies using 3D printing to ramp up production are GE (jet engines, medical devices, and home appliance parts), Lockheed Martin and Boeing (aerospace and defence), Aurora Flight Sciences (unmanned aerial vehicles), Invisalign (dental devices), Google (consumer electronics), and the Dutch company LUXeXcel (lenses for light-emitting diodes, or LEDs).[13]

This will be a theme explored more during the course of this book, but suffice it to say here that the involvement of existing industrial players in developing and using 3D printing, either internally or in partnership with large 3D printing firms such as Stratasys and 3D Systems, would go some way to dispelling the idea that 3D printing is a socially revolutionary technology. Furthermore, it is not only private sector incumbents which have embraced 3D printing: nation-states have also been considering the opportunities and threats 3D printing presents to their activities.[14]

---

[12] See, for example, Trimensional, http://www.trimensional.com/, accessed 10 September 2015; ETH Zurich, 'Transform your smartphone into a mobile 3D scanner', http://www.inf.ethz.ch/news-and-events/spotlights/mobile_3dscanner.html, accessed 10 September 2015.

[13] R. D'Aveni (2015) 'The 3-D Printing Revolution' *Harvard Business Review*, https://hbr.org/2015/05/the-3-d-printing-revolution, accessed 10 September 2015.

[14] C. Arizmendi, B. Pronk and J. Choi (2014) 'Services No Longer Required? Challenges to the States as the Primary Security Provider in the Age of Digital Fabrication' *Small Wars Journal*; L. Grant (2014), 'Bits to Bullets: Australian Military 3DP's New War-Making Strategies and Tactics' 3D Printing Industry, http://3dprintingindustry.com/2014/07/14/bits-bullets-australian-military-3dps-new-war-making-strategies-tactics/, accessed 5 September 2015; J. M. Pearce and A. S. Hasselhuhn (2015),

THE REVOLUTIONARY PROMISE OF 3D PRINTING

3D printing's entry into the mainstream in the developed world has provoked strong reactions. Various sources have hailed 3D printing as a Third or New Industrial Revolution,[15] or as forming part of such a Revolution along with other developments such as the Internet and renewable energy.[16]

Of course, in a technological sense, 3D printing (or rapid prototyping/additive manufacturing) is obviously revolutionary, as encompassing a paradigm shift from traditional subtractive manufacturing. 3D printing enables objects to be manufactured that may have been impossible to create using traditional techniques. The possibilities of manufacturing objects in a quicker and more cost-effective fashion while reducing wastage are other elements of 3D printing which can be seen as technologically revolutionary.

However, the idea that 3D printing is revolutionary is not confined to its technical aspects. Indeed, there are strong currents running through the literature around 3D printing that it is socially and economically transformative, and, in turn, poses significant challenges to the effective enforcement of law. Interestingly, these themes bear a striking resemblance to the initial commentary around the Internet and World Wide Web during the 1990s, as will be explored here.

3D Printing as the End of Scarcity: Atoms Also Want to Be Free

A strong current running through discussion of 3D printing is that it is socially transformative in a way which will ensure we have an abundance of information about how to make complicated objects as well as the means of production being within reach of many more people than previously. In this sense, 3D printing is hailed as bringing society into a 'post-scarcity' age.

There are strong parallels between this rhetoric and similar comments made in response to the rise of the Internet and personal computing. In this case, the ability to create, copy, and disseminate digital information

'Intellectual Property as a Strategic National Industrial Weapon: the Case of 3D Printing' *Engineer: The Professional Bulletin of Army Engineers*, 45(2), 29–31.

[15] The Economist (2012) 'The third industrial revolution', http://www.economist.com/node/21553017, accessed 10 September 2015; C. Anderson (2012) *Makers: The New Industrial Revolution* (New York: Crown Business).

[16] J. Rifkin (2014) *The Zero Marginal Cost Society: The internet of things, the collaborative commons, and the eclipse of capitalism* (Basingstoke and New York: Palgrave Macmillan).

using these technologies contributed to the sentiment that 'information wants to be free'. 'Free' in the English language is ambiguous inasmuch as it can mean 'for no cost' (*gratis*) and 'with little or no restriction' (*libre*). The two meanings of 'free' have been an important distinction for the free software and Creative Commons movements built up around these technological developments in opposition to the restrictive practices envisaged by traditional IP, especially those that would interfere with the freedom (*libre*) to modify and tinker.

Yet the idea of information wanting to be free as in *gratis* as a result of personal computing and the Internet has also permeated some of the 1990s scholarship about the Internet. Notably, DeLong and Froomkin explored the economic impact of this 'freeing' of information via the Internet and personal computing, considering that digitised information exhibited 'different' features to those pertaining to physical objects:

- information is not excludable: owners of information goods and services cannot easily and cheaply exclude others from using or enjoying these commodities since digital data is cheap and easy to copy;
- information is not rivalrous: two can 'consume' the information good as cheaply as one; and
- information is not transparent: 'much of the value added in the data-processing and data-communications industries today comes from complicated and evolving systems of information provision' such that it is not clear to the consumer what precisely they are buying.[17]

Of these features, most relevant to the idea of post-scarcity are information being non-excludable and non-rivalrous. This would suggest that information is in abundance and it is difficult if not impossible to stop others from accessing it, as well as there being no or minimal cost in the production and access. While DeLong and Froomkin acknowledge that there may be attempts to reintroduce excludability and rivalry through technical protection measures (TPMs) and digital rights management (DRMs), they consider such measures to be imposing a (largely unwanted) artificial scarcity.

As mentioned above, similar rhetoric is present regarding 3D printing. Weinberg considers that 3D printing is a 'democratizing' technology

[17] J. B. DeLong and A. M. Froomkin (2000) 'Speculative Microeconomics for Tomorrow's Economy' *First Monday*, 5(2).

which 'may make the creation of physical objects nearby as widespread as the creation of [digitised] copyright-protectable works'.[18] Lemley also identifies 3D printing as one of various technologies, including the Internet as a predecessor and alongside synthetic biology and robotics, which 'radically reduce the cost of production and distribution of things', contributing towards a 'not-too-distant world in which most things that people want can be downloaded and created on site for very little money – essentially the cost of raw materials'.[19]

Yet raw materials themselves and the energy sources to power 3D printers are also subject to other trends which may make them post-scarcity as well. Markus Kayser, a German designer, has created a solar-powered 3D printer which uses sand as a raw material to print glass objects.[20] Another initiative is that of the Perpetual Plastic Factory, which uses plastic glasses as raw material for a 3D printer.[21] If the RepRap is added to this group of developments, then individuals also have the means of building their own 3D printers. Furthermore, there is a strong 'sharing' ethic in the hacker/maker communities which have grown up around low-cost 3D printing and the RepRap, whereby tools, know-how, and other resources are shared with each other, either in person in hacker/makerspaces and Fab Labs,[22] or virtually through sites such as Thingiverse which encourage users to license their 3D printing design files using Creative Commons and free software licences.[23]

Thus, technically speaking, we are increasingly in a post-scarcity society where objects are theoretically as abundant as information became as a result of personal computing and the Internet, and waste is also reduced.

[18] M. Weinberg (2010) *It Will Be Awesome If They Don't Screw It Up: 3D Printing, Intellectual Property, and the Fight Over the Next Great Disruptive Technology*, Public Knowledge White Paper.

[19] M. Lemley (2014) 'IP in a World Without Scarcity' Stanford Public Law Working Paper No. 2413974, p. 2.

[20] M. Kayser (2011) 'The Solar Sinter' http://www.dezeen.com/2011/06/28/the-solar-sinter-by-markus-kayser/, accessed 10 September 2015.

[21] Better Future Factory (2012) 'Perpetual Plastic Project' http://www.betterfuturefactory.com/work/perpetual-plastic-project-ppp, accessed 10 September 2015.

[22] P. Wolf, P. Troxler, P. Y. Kocher, J. Harboe and U. Gaudenz (2014) 'Sharing is Sparing: Open Knowledge Sharing in Fab Labs' *Journal of Peer Production,* Issue #5 Shared Machine Shops.

[23] J. Moilanen, A. Daly, R. Lobato and D. Allen (2015) 'Cultures of Sharing in 3D Printing: What Can We Learn from the Licence Choices of Thingiverse Users?' *Journal of Peer Production,* Issue #6 Disruption and the Law.

In response to these developments, Soderberg and Daoud have considered the implications of physical objects (or at least the ability to make such objects) being 'democratised', and consider that critiques of IP (especially copyright and patents) which arose due to the properties of digitised data in the wake of personal computing and the Internet should also be expanded to tangible property.[24]

3D Printing as the End of Control

Another strong current running through initial Internet literature, and more recently regarding 3D printing, is the idea that these technological developments have resulted in a situation whereby power and authority cannot (and should not) be asserted in a controlling way over individuals' activities, particularly if that power is governmental.

In the early 1990s, cyberlibertarian manifestos denied the ability and authority of the nation-state to control the Internet,[25] and believed they were seeing a 'freeing' of culture and information in the online environment. Due to factors which mostly concerned the content of what was being placed on the Internet (such as the lack of restrictions on what kind of information could be uploaded/downloaded to/from the Internet), its seemingly transnational nature, the lack of de facto government control over the medium (at least the layers of it which were more 'visible' to users), and the initial lack of prominence of large corporate entities at these layers more visible to users (or at least the absence of them acting in a way which impeded users seeing and doing what they wanted on the Internet), it appeared that the Internet represented an autonomous space in which users had control over their actions and online destiny (or at least more control as compared to previous mass mediums such as television or the press).

The most prominent of these manifestos are John Perry Barlow's *A Declaration of the Independence of Cyberspace*,[26] which denied the sovereignty of nation-states over the Internet, and asserted the ability of the Internet community to self-govern, as well as defining the Internet as the

---

[24] J. Söderberg and A. Daoud (2012) 'Atoms Want to Be Free Too! Expanding the Critique of Intellectual Property to Physical Goods' *Triple C Communications, Capitalism & Critique* 10(1).

[25] J. P. Barlow (1996) 'A Declaration of the Independence of Cyberspace' https://projects.eff.org/~barlow/Declaration-Final.html, accessed 10 September 2015.

[26] J. P. Barlow (1996) 'A Declaration of the Independence of Cyberspace'.

place where 'whatever the human mind may create can be reproduced and distributed, infinitely at no cost', thereby claiming the Internet's capacity to collect and disseminate to a potential mass global audience any and all ideas. The somewhat less utopian article *Cyberspace and the American Dream: A Magna Carta for the Knowledge Age* of Dyson et al. nevertheless proclaimed the death of 'bureaucratic' (governmental) power and the 'demassifying' or 'freeing' of institutions and culture (given financial costs were being driven towards zero in cyberspace),[27] which would implicate a lack of necessity for economic regulation and oversight as well.

Similar rhetoric can be seen in literature around 3D printing, particularly when it comes to the possibility of 3D printed weapons (discussed in greater detail in Chap. 3). Suffice it to say here that 3D printing is considered to be one technology, cross-fertilising with other developments, that render the nation-state less potent in regulating individuals' conduct:

> three disruptive technologies—file-sharing, 3D printing, and distributed digital currency—have severely undermined the legal and regulatory capacity of the state, resulting in an anarchic environment where actors' behaviour is determined primarily by factors other than legislation or governmental authority.[28]

This 'anarchic environment' seems to result from the disintermediation from traditional points of control that 3D printing engenders, given the ability to print many kinds of complex objects including dangerous ones in the home. The nation-state is thus less able to control individuals' access to these items and in consequence its own power and authority is diminished.

Technodeterminism

Be that as it may, these themes of 3D printing driving a post-scarcity and post-control society smack of technological determinism, or the idea that technology drives socio-economic development. As can be seen from the

---

[27] E. Dyson, G. Gilder, G. Keyworth and A. Toffler (1994) 'Cyberspace and the American Dream: A Magna Carta for the Knowledge Age' http://www.pff.org/issues-pubs/futureinsights/fi1.2magnacarta.html, accessed 10 September 2015.

[28] G. J. Michael (2013) 'Anarchy and Property Rights in the Virtual World: How Disruptive Technologies Undermine the State and Ensure that the Virtual World Remains a "Wild West"', SSRN Working Paper, http://papers.ssrn.com/sol3/papers.cfm?abstract_id=2233374, accessed 10 September 2015.

quotes above, technodeterminism seems to pervade much of the early scholarship on the Internet and now seems attached to the scholarship on 3D printing, that both of these developments are bringing about a post-scarcity and post-control society.

It is true that technological developments including the personal computer and Internet empower consumers, and give them a much greater productive capacity. The many-to-many nature of Internet communications (as opposed to the one-to-one nature of the telephone, or the one-to-many nature of broadcast and print media) and the very low cost of creating, copying, and disseminating data via the Internet (once Internet access and equipment has been bought) has given rise to the 'prosumer', that is, individuals with the capacity to create online as well as consume the creations of others.[29] The rise of the prosumer also contributed to the phenomenon of 'commons-based peer production': individuals on a decentralised basis collaborating together to produce information and cultural outputs over which no traditional intellectual property right (IPR) is asserted and so the product is free to access and use.[30] This can be conceptualised as a radical alternative to traditional forms of centralisation in the form of the State or centralisation in the form of the firm: a kind of loosely defined 'third way' of organising production.

However, in practice, the picture is more complicated. Firstly, few truly common-based peer production initiatives in the Internet context actually exist. Furthermore, commons-based peer production, as well as other Internet activities, depend on arrangements in the 'physical' world which itself continues mostly to be based on property rights rather than commoning arrangements,[31] 3D printing notwithstanding.

The Internet itself has also seen the emergence of strong axes of power, whether in the form of concentrations of private economic power in Internet markets,[32] in the form of the nation-state,[33] or in an unholy

---

[29] Y. Benkler (2000) 'From Consumers to Users: Shifting the Deeper Structures of Regulations Towards Sustainable Commons and User Access' *Federal Communications Law Journal*, 52, 561–579.

[30] Y. Benkler (2006) *The Wealth of Networks: How Social Production Transforms Markets and Freedom* (New Haven: Yale University Press).

[31] J. M. Pedersen (2010) 'Conclusion: Property and the Politics of Commoning' *The Commoner*, 14, 287–294.

[32] A. Daly (2015) *Mind the Gap: Private Power, Online Information Flows and EU Law* (PhD thesis, European University Institute).

[33] J. L. Goldsmith and T. Wu (2006) *Who Controls the Internet? Illusions of a Borderless World* (Oxford: Oxford University Press).

alliance between the two.[34] This is laid bare in Edward Snowden's revelations about the US National Security Agency and its partners which co-opt large and pervasive private entities such as Google to monitor their users' conduct for the State's benefit, with more contextualisation provided by collecting data about users from Google's myriad products and services.[35] In this sense, it is hard to argue that the Internet is a post-control technology, as both large corporations and nation-states are making use of the Internet for their own ends.

As regards the post-scarcity arguments about the Internet, the late 1990s onwards saw the re-assertion of IPRs over both software and content being shared online, and corporate lobbying for their increased enforcement in cyberspace—a process which has been described by Boyle as 'the second enclosure movement'.[36] The World Intellectual Property Organization (WIPO) Copyright Treaty from 1996 and its implementations in the Digital Millennium Copyright Act (DMCA) in the USA and via a series of EU Directives are examples of this attempt to 'beef up' IPRs and their enforcement when faced with disruptive new information technologies, for the benefit of large corporate copyright holders.[37]

3D printing's trajectory as a purportedly post-control and post-scarcity technology remains to be seen. However, initial observations would point to some similarities with the path the Internet followed, and some differences. As far as similarities go, it would seem that the 'mainstream' experience of both the Internet and 3D printing are converging. As Zittrain noted, many people are using more 'closed' and less 'generative' devices to access the Internet such as smartphones and tablets, which limit user or 'prosumer' innovation but provide a 'safer' and more controlled experience.[38] Similar trends can be observed with 3D printing, where the large

---

[34] M. D. Birnhack and N. Elkin-Koren (2003) 'The Invisible Handshake: The Reemergence of the State in the Digital Environment' *Virginia Journal of Law & Technology*, 8 (6), 1–57; J. Cohen (2012), *Configuring the Networked Self: Law, Code and Everyday Practice* (Yale University Press), p. 177.

[35] See: D. Lyon (2014) 'Surveillance, Snowden and Big Data: Capacities, consequences, critique' *Big Data and Society*, 1–13.

[36] J. Boyle (2003) 'The Second Enclosure Movement and the Construction of the Public Domain' *Law and Contemporary Problems*, 66, 33–74.

[37] See: A. Daly and B. Farrand (2015) 'SABAM v Scarlet: evidence of an emerging backlash against corporate copyright lobbies in Europe?' in D DeVoss and M Rife (eds.), *Cultures of Copyright* (New York: Peter Lang).

[38] J. Zittrain (2008) *The Future of the Internet and How to Stop It* (New Haven: Yale University Press).

3D printer manufacturers offer similarly safe, consumer-friendly yet controlled experiences. Arguably the 'mainstream' experience of these technologies is closer to the 'consumer' paradigm rather than the 'prosumer' paradigm.

Yet there are still some anarchic elements around the edges of both the Internet and 3D printing. The 'Dark Web' content that exists on networks that overlay the public Internet and requires specific software, configurations, or authorisation to access includes peer-to-peer networks such as Tor which serve as a means of accessing content, information, and services that may be at least controversial and at worst illegal.[39] The existence of the RepRap project ensures that individuals are capable of constructing their own 3D printer 'off the radar', which would frustrate attempts to regulate intermediaries such as 3D printing manufacturers, raw materials suppliers, or file-sharing sites in order to ensure that dangerous items cannot be printed out. The social disadvantages of outlawing the RepRap would likely outweigh any advantages in ensuring the effective enforcement of, for example, weapons laws, especially since open projects such as the RepRap enable the technology to be more accessible to underprivileged communities which may not be able to afford 'off-the-shelf' models such as those sold by the major 3D printer manufacturers.[40] As will be seen in Chap. 3, despite (largely successful) efforts to remove the 3D printing gun blueprints from public accessibility, they were still available in some of these less-salubrious parts of the Dark Web.

Thus, in practice, both the Internet and likely also 3D printing do exhibit elements which frustrate the effective control of the technology by centralised forces such as the nation-state and large private sector actors, but the reality is not one of total chaos either. The decentralised nature of both technologies for information and object production does provide challenges to the effective enforcement of the law, as well as raising questions about the extent to which existing categories of law and the assumptions which underpin them are appropriate to the (potential) changes that these technologies bring.

---

[39] M. Ward (2014) 'Tor's most visited hidden sites host child abuse images' BBC News, http://www.bbc.com/news/technology-30637010, accessed 11 September 2015.

[40] S. Dodson (2008) 'The machine that copies itself' *The Guardian*, http://www.theguardian.com/technology/2008/jul/03/copy.machine.reprap, accessed 2 September 2015; J. M. Pearce (2015) 'Applications of Open Source 3-D Printing on Small Farms' *Organic Farming*, 1(1), 19–35.

Arguably difficulties in enforcing laws are not novel, although perhaps seem so in what appear to be highly regulated Western jurisdictions such as those under consideration in this book, despite the widespread existence of 'shadow' or 'informal economies'.[41] One can think of the widespread availability of illegal drugs in such countries as an obvious example of how the 'law in the books' is not always perfectly enforced on the streets.[42] Furthermore, one can also look at economies which are more decentralised in being made up of small and medium businesses rather than large consolidated enterprises where the costs of the state administering, for example the taxation regime, may be greater, and efficacy less, as a result of having to take account of the activities of more enterprises. Thus, it may be argued that it is more difficult to enforce laws in a decentralised economy/society—but not impossible. Rather than following the technodeterministic claims made in the literature mentioned above, it is more expedient to examine what is happening in practice with 3D printing and the various areas of law identified in this book—a task the following chapters aim to do.

OUTLINE OF THIS BOOK

The following three chapters of this book each look at different aspects of 3D printing which give rise to legal issues. The relevant laws and how they apply to 3D printing are explained, along with a snapshot of how legal disputes are playing out in practice.

Chapter 2, the first substantive chapter, will examine how IP is interacting with 3D printing, both at a conceptual level, and then at an empirical level. The relationship between IP and 3D printing has captured the academic imagination, especially given the battles waged between the incumbent content industry and file-sharing services over copyright on the Internet. The extent to which similar conflicts are emerging with 3D printing will be discussed.

---

[41] See: A. Portes, M. Castells and L. A. Benton (eds.) (1989) *The Informal Economy: Studies in Advanced and Less Developed Countries* (Baltimore: John Hopkins University Press); E. L. Feige (ed.) (2007) The underground economies: Tax evasion and information distortion (Cambridge: Cambridge University Press). In the media sector, see: R. Lobato (2012) *Shadow Economies of Cinema: Mapping Informal Film Distribution* (Basingstoke and New York: Palgrave Macmillan).

[42] See: L. Mazerolle, D. Soole and S. Rombouts (2007) 'Drugs Law Enforcement: A Review of the Evaluation Literature' *Police Quarterly*, 10(2), 115–153.

The following chapter, Chap. 3, will address the other major legal controversy that has accompanied 3D printing: the possibility of creating dangerous or defective objects. This has been amplified by the emergence of the 3D printed gun, the Liberator, and the contentious discussions and legal action which have accompanied it. Yet there are also more mundane items that can be printed by 3D printers which may also cause injury or harm, and these are also considered in the chapter.

Then, Chap. 4 examines the emergence of 3D scanning as a companion technology to 3D printing. Similar legal issues arise from 3D scanning to those already examined, especially concerning IP. In addition, given a prominent application of 3D scanning vis-à-vis consumers has been to scan human bodies for fashion and health reasons, data privacy laws are also explored in this chapter.

Finally, some conclusions will be drawn in Chap. 5 on the socio-legal aspects of 3D printing, based on the discussion in the preceding chapters, as well as giving some views on the post-control and post-scarcity arguments discussed earlier in this chapter.

CHAPTER 2

'You Wouldn't Download a Car': 3D Printing and Intellectual Property

Abstract Prominent in discussions about the interaction of law and 3D printing has been the effect that 3D printing may have on IP, in terms of how and when new IPRs are created by the 3D printing process, and how and when the IP of others may be infringed. Given IP disputes, especially around file-sharing, have been one of the defining features of cyberlaw literature and jurisprudence, there is great anticipation about whether similar battles will be witnessed with 3D printing. However, while copyright was mainly at issue in the Internet context, 3D printing also implicates other areas of IP, notably patents, design rights, and trade marks, particularly given the fact that 3D objects are created by 3D printing. Indeed, the phrase 'you wouldn't download a car' from a 2004 Motion Picture Association of America (MPAA) campaign aimed at the illicit sharing of copyrighted items takes on new dimensions in the 3D printing context, especially since 3D printed cars have been developed— and it is already possible to download more mundane 3D printing files for car parts. This chapter explores this interaction between 3D printing and IP, both theoretically and practically, looking at how this relationship is playing out so far.

© The Editor(s) (if applicable) and The Author(s) 2016   19
A. Daly, *Socio-Legal Aspects of the 3D Printing Revolution*,
DOI 10.1057/978-1-137-51556-8_2

This chapter, the beginning of the substantive section of the book, will commence with a consideration of 3D printing's implications for IP law. How, when, and where the different types of IP—copyright, patents, trade marks, and design rights—apply to the process of 3D printing will be examined, as well as the potential for their infringement.

Disputes that have already arisen over IP and 3D printing will be detailed, which hitherto have mainly involved copyright infringement claims and US DMCA takedown notices relating to 3D printing design files being uploaded to file-sharing platforms such as Thingiverse. However, unlike the situation with predecessor technologies such as the Internet, IP battles in the 3D printing sphere have also concerned alleged infringements by major players of creative and inventive works belonging to individuals, reversing the direction of traditional 'piracy' claims.

Prospects for the enforceability of IP will be explored, such as DRMs and other TPMs being used on 3D printing design files and/or 3D printing machines. Comparisons will be drawn with the success (or not) of these measures in attempting to stem IP infringements in other technologies, alongside the disadvantages they carry for users' ability to engage in legal conduct, as well as a potential lack of interoperability with rivals' printers and software, thus bringing up issues of competition and dominance which will also be discussed.

A Brief Introduction to Intellectual Property

IP is the area of law which has played the most prominent role in discussions related to 3D printing, until the development of the 3D printed gun discussed in the next chapter. IP protects intangible assets, and is subdivided into various different rights protecting different subject matter. Broadly speaking, the purpose of contemporary IP is to encourage creation and innovation through the award of exclusive rights over such creations and innovations for a certain period of time for creators/inventors or those to whom they have transferred their rights. This can be seen, for instance, in the 'Copyright Clause' of the US Constitution:

> To promote the Progress of Science and useful Arts, by securing for lim-
> ited Times to Authors and Inventors the exclusive Right to their respective
> Writings and Discoveries.[1]

Whether the grant of IPRs actually does promote creativity and innovation is contested. There is little empirical evidence to suggest that
this happens in practice.[2] The rise of commons-based ownership and
management of what otherwise would be restricted by exclusive rights,
through Creative Commons licensing and the free software movement,
also challenges the assumptions on which IP rests.[3] Nevertheless, despite
the doubt over IP's effectiveness in achieving its goals, it is very much
an area of law still in force, and applicable to 3D printing activities.

Indeed, unlike the Internet context, in which mainly copyright over
digitised content files has been implicated, 3D printing involves all main categories of IP: copyright, patents, trade marks, and design rights. This brings
certain theoretical complications, which will be explained later in more detail.

The main categories of IP relevant to 3D printing are as follows:

- *Copyright* is the part of IP which protects certain expressions of
  ideas (rather than the ideas themselves), with 'traditional' types of
  copyrighted material being literary, dramatic, musical, or artistic
  works. Copyright in jurisdictions such as the USA, Australia, and
  EU Member States does not have to be registered in order to come
  into existence.
- *Patents* protect new and inventive industrial products and processes.
  Inventors must register their inventions in order to be granted a
  patent.

---

[1] US Constitution, Article 1, Section 8, Clause 8. Other legal traditions, such as that of
civil law countries including France, have placed more emphasis on authorship as constitu-
tive of IPRs, and less influence on IP as being instrumental to innovation. See: C. Chinni
(1992) 'Droit d'Auteur versus the Economics of Copyright: Implications for American
Law of Accession to the Berne Convention' *Western New England Law Review*, 14(2),
145–174.

[2] M. Boldrin and D. K. Levine (2008) *Against Intellectual Monopoly* (Cambridge:
Cambridge University Press).

[3] E. G. Coleman (2012) *Coding Freedom: The Ethics and Aesthetics of Hacking* (Princeton:
Princeton University Press).

- *Trade marks* are recognisable signs which distinguish products or services as coming from a particular source. Whether trade marks must be registered to be effective depends on the jurisdiction.
- *Design rights* protect the visual appearance (the look or shape) of a product which is not purely utilitarian. In the UK, design rights exist in registered and unregistered forms. In other jurisdictions, such as Australia, design rights must be registered, and in the USA, registered design rights are called design patents.

The action of 'passing off' also features in common law systems such as England and Australia to protect the reputation a particular individual or company has in a distinctive designation or get-up. This is known as 'misappropriation' in the USA, part of the tort of unfair competition, and trade dress.

There is some level of global harmonisation of IP law, through international treaties. Early treaties which are still relevant today are the Paris Convention for the Protection of Industrial Property (mainly concerning patents and trade marks) and the Berne Convention for the Protection of Literary and Artistic Works (copyright), both administered by the WIPO. There is also the World Trade Organization-administered Trade-Related Aspects of Intellectual Property Rights (TRIPS) Agreement, which is more recent than the aforementioned WIPO treaties. In addition, trade agreements have been another vehicle through which some more globalised standards of IP protection and enforcement are emerging—this was attempted with the (ultimately ill-fated) Anti-Counterfeiting Trade Agreement (ACTA), and may come to fruition in the Transatlantic Trade and Investment Partnership (TTIP) currently under negotiation between the EU and the USA, and the Trans-Pacific Partnership (TPP) treaties, which was recently signed by representatives of various Pacific Rim countries, including the USA and Australia. The EU itself, in its pursuit of regional harmonisation and the creation of an internal Single Market, has entered the fore with various laws which concern IP throughout the Union.

Despite these harmonisation attempts, precisely what is protected by laws on copyright, patents, and so on—and exceptions to the exclusive rights granted by such protection—varies from jurisdiction to jurisdiction. This is significant for 3D printing, as a transnational phenomenon whose design files are disseminated—and printing machines sent—across borders, as will be explored in more detail below.

Intellectual Property's Intersection with 3D Printing

There are various points in the 3D printing process at which IP may be created (or infringed):

1. 3D printing design (computer-aided design—CAD) file (code)—and the software with which it interacts
2. the substance within the design file, that is, the 'artistic' creation to be printed
3. the final 3D printed object
4. the online repositories where design files are uploaded and shared—such as Thingiverse.

Precisely how the different kinds of IP protection will apply at these different points in the 3D printing process will ultimately depend upon the individual circumstances at hand, as will be illustrated further in the following sections. Suffice it to say here that it is hard to lay down hard and fast rules around the application of IP law(s) to 3D printing.

This section will analyse how IP interacts with 3D printing at the first three points listed above.

Copyright

Copyright, as mentioned above, is an IPR which does not need to be registered with any authority for it to come into existence. While copyright has traditionally protected literary and artistic works, the scope of copyright protection has been greatly expanded in recent decades—for instance, in many jurisdictions, software code can now be subject to copyright protection.

Of the elements of the 3D printing process, all three may encompass a point at which copyright subsists:

- The 3D printed design file's code may attract copyright protection, in light of the expansion of copyright's categories as just mentioned.
- The design contained within the file is likely to attract protection as an artistic work.
- The physical 3D object to be printed may, in certain circumstances, also attract copyright protection as an artistic work.

Computer-Aided Design File

From a European and British perspective, Mendis has discussed the question of whether the CAD file itself is capable of being protected as a copyright work.[4] The EU Software Directive, which extends copyright protection to computer programs,[5] does so by protecting the expression of computer code and not the functionality of the software,[6] and recent case law on copyright from the Court of Justice of the EU (CJEU) has emphasised the extent to which a work should demonstrate the author or creator's 'own intellectual creation'.[7] Mendis considers that this case law, and its discussion in domestic British courts, has left open the question of whether CAD files can be protected by copyright.[8] Nevertheless, in domestic UK law, the category of literary works includes computer programs and 'the preparatory design material' for a computer program,[9] and it would seem that a 3D printing design file would fall within this definition and thus be protected by copyright in the UK.[10] This is supported by some comments made by Laddie J in *Autospin v Beehive* before the advent of 3D printing that such a file would be protected by copyright, that is '[a] literary work consisting of computer code ... represent[ing] the three dimensional article'.[11]

The situation in the USA is also unclear, but the academic and judicial discussion does not seem inclined to view CAD files as being protected by copyright, particularly due to the creative–functional distinction in American IP law. This is discussed in more detail in the next subsection, but suffice it to say here that, broadly speaking, creative works are capable of being protected by copyright and functional works are capable of being protected by patents. It is true that US law, similar to that in the EU, has recognised for some time that software—both source and object

---

[4] D. Mendis (2014) '"Clone Wars": Episode II – The Next Generation: The Copyright Implications Related to 3D Printing and Computer-Aided Design (CAD) Files' *Law, Innovation and Technology*, 6(2), 265–281.

[5] Council Directive 2009/24/EC of 23 April 2009 on the legal protection of computer programs (codified version) [2009] OJ L111/16.

[6] Mendis (2014) '"Clone Wars II": Episode II', 270.

[7] For example: Case C-5/08 *Infopaq International A/S v Danske Dagblades Forening* [2010] ECR I-6569.

[8] Mendis (2014) '"Clone Wars II": Episode II', 271.

[9] Copyright, Designs and Patents Act 1988, s 3(1)(b) and (c).

[10] S. Bradshaw, A. Bowyer and P. Haufe (2010) 'The Intellectual Property Implications of Low-Cost 3D Printing' SCRIPTed, 7(1), 5–31, 24.

[11] *Autospin (Oil Seals) Ltd v Beehive Spinning (A Firm)* [1995] RPC 683, 698.

code—is capable of being protected by copyright, as well as being patentable, although this has not been without controversy.[12]

Simon, on examining US statute and case law on copyright—in particular that copyright protection cannot extend to 'any idea, procedure, process, system, method of operation, concept, principle, or discovery, regardless of the form in which it is described, explained, illustrated, or embodied' (17 U.S.C. § 102(b)), since these are items falling under the domain of patent law—considers that since 3D design files are dictated by functional considerations (their function being a set of instructions for 3D printing), then they ought not to enjoy copyright protection (or at least there should be a presumption against them enjoying copyright protection).[13] In addition, US law on software copyright puts forward the requirement that the code must be the 'writing of an author'. It would seem that the code contained within 3D printing design files is not analogous to computer software which can be protected by copyright as it cannot be considered in any way to be the 'writing of an author': unlike computer source code, the design file's code is never in a human-readable form. As Simon puts it, '3D design files and computer software are electronic data stored in computer memory to be used with the assistance of a machine as provided for in § 102(a)'.[14]

Thus, already it can be seen that there may be vastly differing views in only the UK and USA as to whether the CAD file itself can be protected by copyright as a literary work, despite the aforementioned attempts at global harmonisation of IP law. In any event, even if the British approach is adopted, it would seem that that copyright would only be infringed if there is actual copying of a 'substantial part' of the file's code—rather than a copy being made of the design contained within that file.

Design for the Eventual Object

Whether copyright protects the design for the eventual 3D printed object contained within the file is also problematic. According to Weinberg, the US law position on this is designs are only protected by copyright to the

---

[12] P. Samuelson (1988) 'American Software Copyright Law' *Columbia-VLA Journal of Law & the Arts*, Vol. 13, 61–75.

[13] M. Simon (2013) 'When Copyright Can Kill: How 3D Printers are Breaking the Barriers between "Intellectual" Property and the Physical World' *Pace Intellectual Property, Sports and Entertainment Law Forum*, 3(1), 59–97, 71.

[14] Simon (2013) 'When Copyright Can Kill', 79.

extent that they go beyond the utilitarian requirements of designing a useful article.[15] Weinberg also considers that if the file contains the design for a creative object, then that is protected by copyright and so creating that object in physical form using, for example, a 3D printer would require permission from the copyright holder since the physical object is a derivative work of the design.[16] The important element here is whether the design is for a *creative* object or some other kind of object—US law draws a distinction between creative and functional objects, with the former attracting copyright protection and the latter attracting potential patent protection.

Yet there are scenarios in which a given object may have both creative and functional attributes, in which case US law employs the severability test, by which any decorative elements of the object that exist beyond the scope of the useful object can be protected under copyright.[17] As Weinberg notes, there is no simple straightforward severability test—it remains a fact-finding inquiry. Thus, it will depend on the circumstances at hand whether the design is for a creative or functional object, and, accordingly, whether it is protected by copyright. Simon considers that 3D design files may be protected under US copyright law as 'technical, mechanical, engineering, or architectural drawings',[18] but importantly, while copyright will protect the document itself against being copied, it will not protect the design portrayed within the document.[19] Thus, 3D printing files protected in this way do not permit the copyright holder to 'prevent a third party from creating a utilitarian object based upon the drawing, in cases where no unauthorized reproductions of the drawing are subsequently used in creating the copied utilitarian object'.[20]

In UK law, the graphic design contained within a CAD file is likely to constitute an artistic work in the form of 'a graphic work, photograph, sculpture or collage, irrespective of artistic quality'.[21] However, it will also

[15] Weinberg (2010) *It Will Be Awesome If They Don't Screw It Up*.

[16] M. Weinberg (2013) *What's The Deal With Copyright and 3D Printing*, Public Knowledge White Paper, p. 19.

[17] Weinberg (2010) *It Will Be Awesome If They Don't Screw It Up*.

[18] Simon (2013) 'When Copyright Can Kill', 82.

[19] U.S. Copyright Office (2015) 'Copyright Registration for Works of the Visual Arts', Circular 40, www.copyright.gov/circs/circ40.pdf, accessed 11 September 2015.

[20] Simon (2013) 'When Copyright Can Kill', 83.

[21] Copyright, Designs and Patents Act 1988, s 4(1)(1).

likely constitute a 'design document', and there is no copyright infringement by actually making the article whose design is contained within such a document, unless the design is for an artistic work (defined below).[22] 'Design document' is defined as 'any record of a design, whether in the form of a drawing, a written description, a photograph, data stored in a computer or otherwise'. While it would seem that the designs would attract copyright protection,[23] making objects from them, by virtue of section 51 of the Copyright, Designs and Patents Act (CDPA), without the copyright owner's permission would not be an infringement, so long as those objects themselves do not constitute artistic works—although given the broad definition of artistic works, Margoni considers that section 51's relevance will be limited.[24] However, unauthorised copying of the file itself would still be a copyright infringement in UK law, and unregistered design rights may also be infringed by making an article from such a file.

In addition, if the design is based on or embodies pre-existing artwork, then it may be an infringement of the copyright in that work.[25] Yet there is another twist to UK law on this point—if an artistic work has already been exploited with the copyright holder's permission via commercial industrial processes, then the work is only protected for 25 years from when it has first been marketed, and so after that 25-year period, it can be copied by making articles of any description without there being an infringement,[26] although the original design would still receive full-term copyright protection.[27] This would seem to entail that making articles with such a design would not constitute a copyright infringement, although making a graphic design file containing the design for such an article may infringe the original design's copyright.

---

[22] Copyright, Designs and Patents Act 1988, s 51.

[23] Bradshaw, Bowyer and Haufe (2010) 'The Intellectual Property Implications of Low-Cost 3D Printing', 24.

[24] T. Margoni (2013) 'Not for Designers: On the Inadequacies of EU Design Law and How to Fix It' *Journal of Intellectual Property, Information Technology and E-Commerce Law*, 4(3) 225–248, 236.

[25] D. Mendis (2013) '"The Clone Wars" – Episode 1: The Rise of 3D Printing and its Implications for Intellectual Property Law – Learning Lessons from the Past?' *European Intellectual Property Review*, 35(5), 155–169, 167.

[26] C Copyright, Designs and Patents Act 1988, s 52.

[27] Bradshaw, Bowyer and Haufe (2010) 'The Intellectual Property Implications of Low-Cost 3D Printing', 23.

3D Printed Object

As for whether the 3D printed object itself will attract copyright protection, in US law a broad category of sculptural works can enjoy copyright protection, but any 'useful article' ('an article having an intrinsic utilitarian function that is not merely to portray the appearance of the article or to convey information') is excluded from this protection.[28] Of course, there will be objects which embody both aesthetic and useful qualities, in which case the severability test applies again, resulting in decorative or aesthetic elements of the object enjoying copyright protection.

In UK law, artistic works attracting copyright protection include 'sculptures' which are protected 'irrespective of artistic quality', and 'works of artistic craftsmanship'. The legislation does not clearly define what both of these categories include, but subsequent case law, and its consideration in *Lucasfilm v Ainsworth*, has defined sculptures as items whose 'intrinsic quality' is 'to be enjoyed as a visual thing', even if it had other uses beyond the aesthetic—but industrial prototypes have been excluded.[29] Works of artistic craftsmanship would seem to be those objects created for artistic purposes, and not for other purposes. For both sculptures and works of artistic craftsmanship, the intention of the creator seems to be key as to whether they receive copyright protection as artistic works. Bradshaw et al. consider that the current situation is that copyright protection 'is confined to objects created principally for their artistic merit', although copyright protection may also attach to any graphic design on the surface of an object, even if the object itself is not an artistic work.[30] However, Mendis has pointed to the recent aforementioned European case law emphasising the 'right' kind of authorial input to establish copyright protection as possibly suggesting a divergence from this existing UK jurisprudence emphasising the artistic quality of the object.[31]

Yet, such 3D printed objects, even if they are obvious candidates for constituting an artistic work, may fall into the trap of being insufficiently original to attract copyright protection in their own right, if they have been constructed based on a 3D printing design file. There is a series of cases in US law which have found 3D costumes based on two-dimensional (2D) designs to be insufficiently original to attract copyright protection in themselves.[32]

---

[28] 17 U.S.C. § 101–02 (2006).

[29] *Lucasfilm Ltd v Ainsworth* [2011] UKSC 39, 118(vii).

[30] Bradshaw, Bowyer and Haufe (2010) 'The Intellectual Property Implications of Low-Cost 3D Printing', 22.

[31] Mendis (2014) '"Clone Wars II": Episode II', 274.

[32] Simon (2013) 'When Copyright Can Kill', 86–87.

Another trap for such objects is that they may constitute 3D representations of 2D items which are protected by copyright. This has been established in UK law as potentially infringing copyright, if the 2D item is not a design document, although the prior commercial exploitation of such items may render their term of copyright protection much shorter.[33]

Patents

While there will be less subjects for patent protection in the 3D printing process, what can be patented will enjoy stronger protection than what copyright offers. Unlike copyright, an application must be made for a patent, and the subject matter of the patent request must fulfil certain criteria before a patent is granted. These requirements include the elements of novelty, non-obviousness (in the USA—'inventive step' in Europe) and usefulness (in the USA—'susceptible of industrial application' in Europe). The elements of novelty and invention require that a patent application be judged against the 'prior art', that is all the inventions already in existence, and the provisional invention must also not be 'obvious' to another person with similar skills in the same technology. Thus, few items ought, in practice, to attract patent protection—and unlike with copyright, it should be clear what is patented and what is not. If a patent is granted, that permits the grantee to make, use, and sell the invention for a period of 20 years in both the EU and USA (as a result of harmonisation via TRIPS), although the grantee also has to abide by a disclosure requirement regarding the information needed to make the invention. Prior to the mass availability of 3D printing, the difficulty in making such an invention (in terms of time, cost, and skill) from this disclosed information was a major barrier to patent infringement.[34]

Unlike copyright, which in requiring an element of 'copying' for infringement to be found permits parallel creations (namely two people separately creating the same thing, independently of each other—although 'unconscious' or 'subconscious' copying may also be infringing),[35] a patent will be

[33] Bradshaw, Bowyer and Haufe (2010) 'The Intellectual Property Implications of Low-Cost 3D Printing', 23.

[34] See: D. H. Brean (2013) 'Asserting Patents to Combat Infringement via 3D Printing: It's No "Use"' *Fordham Intellectual Property, Media & Entertainment Law Journal*, 23, 771, 781–82; D. Desai and G. Magliocca (2013–2014) 'Patents, Meet Napster: 3D Printing and the Digitization of Things' *Georgetown Law Journal*, 102, 1691.

[35] See: *Francis Day Hunter v Bron* [1963] Ch. 587, 612 *per* Willmer LJ; *Bright Tunes Music v. Harrisongs Music* 420 F. Supp. 177 (S.D.N.Y. 1976).

infringed by any unauthorised reproduction of that invention, regardless of whether there is actual copying or of whether the perpetrator is aware of the patent's existence. Weinberg considers that, as a result, 3D printing is likely to increase the numbers of 'innocent' patent infringers—that is, those infringing a patent of whose very existence they are not aware.[36] Nevertheless, this infringement, especially if it is happening in the home, may be extremely difficult to detect. So again, intermediaries may become a target for patent owners. However, there are far fewer patents issued than items subject to copyright protection—and what is protected by patent is protected for a much shorter term than copyright—so there is a low likelihood overall that objects made using a 3D printing infringe a patent. In addition, in many cases, it will be legal in the USA to manufacture replacement parts of a patented item, so long as those individual parts are not themselves covered by a specific patent, and so long as the repair does not constitute a reconstruction of the item in its entirety from constituent parts. Weinberg also considers that replacing parts of a patented item in a way which gives it a new or different functionality will create a new item, and so will not constitute patent infringement in US law.[37] Another difference in US patent law from copyright law is that there is no equivalent of the expansive 'fair use' defence to infringement.

Interestingly, in UK law, there is an exception to patent infringement for private, non-commercial purposes, and another exception for 'experimental purposes relating to the subject-matter of the invention'.[38] It might be considered that much personal or consumer use of 3D printing will fit into either of these categories—private home use or experimental use. However, a provision of UK law which might be considered problematic for 3D printing is that 'making' a patented invention in an unauthorised way includes manufacturing the invention from scratch and this has been interpreted to encompass *inter alia* undertaking such a comprehensive refurbishment of an item so as to remanufacture it in practice.[39] This may make creating spare parts for a patented item using a 3D printer legally tricky, as well as raising the question of whether tweaks, mash-ups,

---

[36] Weinberg (2010) *It Will Be Awesome If They Don't Screw It Up.*

[37] Weinberg (2010) *It Will Be Awesome If They Don't Screw It Up*, p. 10, citing *Husky Injection Moulding System Ltd v R&D Tool and Engineering Company*, 291 F.3d 780 (Fed Cir 2002).

[38] Patents Act 1977, s 60(5)(a) and (b).

[39] *United Wire Ltd v Screen Repair Services (Scotland) Ltd*, [2001] FSR 24.

or remixes of patented items will be viewed as a kind of remanufacture.[40] Additionally, the extent to which prosumer use of others' patents will fit into the category of private and non-commercial use is another important question—at what stage is the line crossed into public and commercial use?

However, intermediary liability for patent infringement has not yet developed to accommodate the changes that digitisation has brought. In US law, 3D printer or scanner manufacturers and the providers of raw materials for 3D printing are unlikely to be held liable for 'contributory infringement' if what they provide is used by someone to infringe a patent since these materials and equipment are 'dual' or 'multi' use—they can be used for both infringing and non-infringing purposes. The situation for file-sharing sites is less clear. Desai and Magliocca note that in the USA, sites such as Napster and eBay were liable for contributory copyright or trade mark infringement only if they had specific knowledge that infringing material was being traded or sold and did not take appropriate action—and that it may take considerable litigation to clarify the position of similar intermediaries vis-à-vis 3D printing files which infringe patents, especially in the absence of an equivalent of the DMCA for patents.[41] In response, they advocate a similar procedure to the DMCA takedown notices being enacted for file-sharing sites hosting 3D printing design files that infringe patents, alongside a limitation of infringement liability for personal use of patented inventions via 3D printing. Weinberg considers that at present, those downloading 3D printing design files which would facilitate the infringement of a patent without further actions would not be engaging in sufficient activity to be found liable for infringement, and it would need to be proved that that individual had actually made the patented object, and this is also true of finding an intermediary liable in a contributory fashion under US law; in addition, there appears to be a requirement that intermediaries have the requisite knowledge or intent to cause someone else to infringe a patent.[42]

In the EU too, indirect patent infringement claims targeting intermediaries may also seem attractive to patent holders due to the difficul-

---

[40] Mendis (2013) '"The Clone Wars" – Episode 1', 160.
[41] Desai and Magliocca (2013–2014) 'Patents Meet Napster'.
[42] Weinberg (2010) *It Will Be Awesome If They Don't Screw It Up*, p. 14.

ties and expense in tracking direct infringers.[43] In UK law, a patent will be infringed by anyone who 'supplies or offers to supply... any of the means relating to an essential element of the invention, for putting the invention into effect'.[44] Case law has held that providing a kit of parts may constitute these 'means'.[45] Bradshaw et al. consider that supplying a 3D printer, raw materials for the printer, and a 3D printing design file might constitute such 'means', but it is certainly not clear that a 3D printing design file alone constitutes 'means'[46]—and presumably, this would be the same of someone merely supplying a 3D printer or the raw materials. However, Mendis considers that a 3D printing design file would constitute 'means'—she considers it an 'essential element' of the invention which puts the invention into effect—and thus disseminating it would constitute infringement.[47] Yet, Bradshaw et al. also question whether a 3D printing design file might be considered as a document describing the patent rather than a source of infringement, and point to this being a subject on which legislative or judicial clarification may be necessary.[48] With the absence of case law on these issues, both in the UK and other European national jurisdictions, it remains to be seen to what extent intermediaries will be liable for indirect patent infringement by disseminating 3D printing design files which contain a patented item.[49]

Designs

Design rights protect the appearance of items, especially those which have an industrial or commercial use (and so may not be protected by other areas of IP). Some jurisdictions such as the UK (and subsequently the whole of the EU) recognise registered and unregistered

---

[43] R. M. Ballardini, M. Norrgård and T. Minssen (2015) 'Enforcing patents in the era of 3D printing' *Journal of Intellectual Property and Practice* 10(11), 850–866.

[44] Patents Act 1977, s 60(2).

[45] *Rotocrop v Genbourne* [1982] FSR 241.

[46] Bradshaw, Bowyer and Haufe (2010) 'The Intellectual Property Implications of Low-Cost 3D Printing', 27.

[47] Mendis (2013) '"The Clone Wars" – Episode 1', 161.

[48] Bradshaw, Bowyer and Haufe (2010) 'The Intellectual Property Implications of Low-Cost 3D Printing', 27.

[49] Ballardini, Norrgård and Minssen (2015) 'Enforcing patents in the era of 3D printing'.

design rights; others such as Australia only protect registered designs,[50] although unregistered design may be protected by other areas of law such as passing off.[51]

Obtaining a design right, similar to other forms of IP, gives the owner a bundle of exclusive rights over the design: to use, sell, license, and protect the design. To register a design, the design must be both new or novel and distinctive or individual: meaning that it is not identical to any design previously disclosed to the public, and it is not substantially similar to any design published anywhere in the world (Australian law)/it produces a different overall impression on a user compared to any other design previously available (UK law).

Design patents in US law are analogous to design rights in other jurisdictions. These protect the novel ornamental, non-functional design of a functional item. Infringement only takes place if the actual design represented in the patent application is copied.

Registered Designs

Registered designs in UK law have been heavily influenced by EU law, and in fact are a subject of continental harmonisation via the Design Directive[52] and the Community Design Regulation.[53] In consequence, design rights can be registered either nationally in the UK, or a designer can opt for EU-wide design protection.

As far as 3D printing is concerned, Bradshaw et al. have considered the interaction between design rights and 3D printing under UK law. In their view, many items which would be attractive to 3D print will not be protected as registered designs in the UK due to a number of 'carve outs' in the law:[54] especially that features of a product dictated solely by technical function may not be protected as registered designs;[55] and the 'must

[50] See: M. Adams (2013) *The 'Third Industrial Revolution': 3D Printing Technology and Australian Designs Law* (Bachelor of Laws (Honours) thesis, Monash University).

[51] *Peter Bodum A/S v DKSH Australia Pty Ltd* (2011) IPR 222. See: A. Scardamaglia (2012) 'Protecting product shapes and features: beyond designs and trade marks in Australia' *Journal of Intellectual Property Law & Practice*, 7 (3), 159–161.

[52] Council Directive 98/71/EC of 13 October 1998 on the legal protection of designs [1998] OJ L289/28.

[53] Council Regulation (EC) 6/2002 of 12 December 2001 on Community Designs [2002] OJ L3/1.

[54] Bradshaw, Bowyer and Haufe (2010) 'The Intellectual Property Implications of Low-Cost 3D Printing', 15.

[55] Registered Designs Act 1949, s 1C (1).

fit' exception (for products whose shape is determined by the need to connect to or fit into or around another product).[56] Furthermore, component parts of a complex product may only be protected as registered designs if they are visible to the user in ordinary use in addition to being of novel and individual design.[57]

Importantly, UK law includes an exception to infringement for registered designs which have been copied 'privately and for purposes which are not commercial'.[58] Thus, home-printing an item for an individual's own use will not infringe any registered design right; while non-commercial uses which are not personal are not covered by this exception, they do enjoy a separate 'fair dealing' exception.[59]

Thus, the position of intermediaries such as file-sharing sites is unclear, particularly the extent to which the 'private' and 'non-commercial' uses can be encompassed by activity intermediated by such sites which involve files that may infringe registered design rights. In addition, whether users using these sites can qualify as 'private' use is unclear.[60]

Unregistered Designs

Unregistered design rights arise in a similar way to copyright protection in UK law, namely automatically upon creation, and similarly are only infringed by actual copying.[61] An unregistered design right can protect the shape and configuration of an object but not its surface decoration or method of construction.[62] In UK law, there is a requirement of originality for such designs (similar to that of copyright law),[63] as well as a 'must fit' exception, and a 'must match' exception analogous to the complex repair provision for registered designs.

As mentioned above, actual copying is required to establish infringement, which can be constituted by making articles to the design or making a design document encompassing the design for the purpose of enabling such items to be made.[64] There is no explicit 'personal and non-commercial' use exception for unregistered design rights, but it seems that

---

[56] Registered Designs Act 1949, s 7A(5).
[57] Registered Designs Act 1949, s 1B(8).
[58] Registered Designs Act 1949, s 7A(2)(a).
[59] Registered Designs Act 1949, s 7A(2)(b).
[60] Mendis (2013) '"The Clone Wars" – Episode 1', 164.
[61] Copyright, Designs and Patents Act 1988, s 226(2).
[62] Copyright, Designs and Patents Act 1988, s 213(2), (3)(a), and 3(c).
[63] Copyright, Designs and Patents Act 1988, s 213(4).
[64] Copyright, Designs and Patents Act, 1988 s 226(1).

the exclusive rights derived from an unregistered design right only relate to commercial use of the design, and so unauthorised non-commercial use is not infringing.[65] Interestingly, this non-commercial use does not have to be private (as is the case for registered design rights, as mentioned above), and so would seem to permit public uses that are non-commercial, for example, by charities and state institutions.

It would seem that sharing design files on file-sharing sites that contained an unauthorised version of the protected design would constitute an infringement of that design right,[66] however if done so for non-commercial purposes, this would seem to fall within the implicit exception to infringement. The position may be different for the file-sharing sites themselves: if it is judged that they profit in some way from the file-sharing activities, then they may be liable for secondary or indirect infringement.[67] Yet proof of copying may be difficult to establish and will depend very much on the actual scenario at hand.

Trade Marks and Passing Off

As mentioned earlier, trade marks are used to identify the source of particular products or services in a way which distinguishes them from similar products and services coming from other sources. In the past, trade marks usually related to graphical signs, brand names, and words. Yet more recently, 'almost anything at all that is capable of carrying meaning' can comprise a trade mark,[68] and this has opened up the categories of potential trade mark protection to colours, scents, sounds, aspects of packaging, and—importantly for 3D printing—2D and 3D shapes.

The strongest way of protecting a trade mark is usually to register it, which gives the owner a bundle of exclusive rights, including using the trade mark as a brand name, authorising others to use the trade mark, and selling the trade mark to another party. While unregistered trade marks can be protected, it is more difficult to do so than if the mark is registered. The scope of trade marks is determined by 'use'—the owner must use the mark vis-à-vis particular goods and services, and may be unable to enforce the trade mark vis-à-vis those goods and services for which the trade mark is not being used. Trade marks must also be capable of distinguishing particular goods and services

---

[65] Copyright, Designs and Patents Act, 1988 s 226(1).
[66] Mendis (2013) '"The Clone Wars" – Episode 1', 165.
[67] Mendis (2013) '"The Clone Wars" – Episode 1', 165.
[68] *Qualitex Co v Jacobson Products Co*, 514 US 159, 162 (1995).

from those of others—if the desired mark is too generic and not sufficiently distinctive, then it will usually not be capable of registration. Another problematic scenario is when someone wishes to register a trade mark which is the same or very similar to a pre-existing mark, or would mislead the public about the nature of the goods or services. While there is a degree of harmonisation of trade mark law under TRIPS, different jurisdictions still have their quirks.

Scardamaglia has examined the interaction of 3D printing and Australian trade mark law.[69] Shapes can be registered as trade marks in Australia assuming they fulfil the general elements of trade mark registration, that is, that it must function as an indicator of source or badge of origin to distinguish one person's goods and services from another.[70] Unlike in US law, there is no functionality doctrine in Australia preventing the registration of functional shapes as trade marks, and instead the issue of whether the proposed trade mark functions as a distinguishing indicator of source or badge of origin is likely to be more important. The protection of product design under the 'trade dress' category of US trade mark law also is contingent on that product acquiring a distinct association with a specific manufacturer, with the result that most product designs will not be protected in this way.[71] Trade dress also cannot protect 'essential features' of the product which are those which would put competitors at a 'significant non-reputational disadvantage' if they were unable to use it, or would affect the cost or quality of the device.[72]

In the US law context, Desai and Magliocca consider that 3D printing may both lead to more trade dress infringement (or unauthorised third-party use of those marks) and may make it difficult to establish trade dress protection for a particular product design in the first place. The latter situation may arise due to 3D printing making it more difficult to show proposed trade dress has been subject to substantially exclusive use by the applicant since individuals could be printing such objects in their houses with a 3D printer.[73]

In addition, the use of a sign which is not confusing for consumers will not infringe the rights of a trade mark owner. Furthermore, unlike in the USA and UK, there is no doctrine of trade mark dilution in Australia,

---

[69] A. Scardamaglia (2015) 'Flashpoints in 3D Printing and Trade Mark Law' *Journal of Law, Information & Science* (forthcoming).

[70] Trade Marks Act 1995, s 17.

[71] *Wal-Mart Stores v Samara Brothers*, 529 US 205, 213–215.

[72] *Traffix Devices v Marketing Displays*, 532 US 23, 33 (2001).

[73] Desai and Magliocca (2013–2014) 'Patents Meet Napster', 1711.

which seeks to protect the prestige of a trade mark and protect against those copying the mark and selling it, even where there is no confusion as to source or origin. Interestingly, Desai and Magliocca remark that any justification for the post-sale confusion doctrine in US law (that the unauthorised use of a trade mark by a third party can injure the trade mark owner even if purchasers of the infringing product are not confused as to origin) is eroded by 3D printing since they believe consumers will have less or even no reason to think that any use of a trade mark observed outside of a retail store was made by the trade mark owner, or that such products being used by someone else were actually made by that person using a 3D printer.[74] Weinberg considers that making something with a trade mark in an individual's own home with a 3D printer ought not be an infringement of that trade mark, but as soon as such an item was sold, even in an informal way, then this would constitute a 'use in commerce' of the trade mark and so be infringing.[75]

In the 3D printing context, trade marks are likely to figure in three scenarios: items being printed which incorporate a graphical 2D trade mark on their surface, 3D representations of 2D trade marks being printed, and 3D items which themselves constitute or replicate a 3D shape mark.

Trade mark infringement may not be readily found in the 3D printing context even if a trade mark is being used. Merely copying a trade mark is insufficient to establish infringement, unlike the situation with copyright and patents. Trade mark law gives the owner the exclusive right to use the sign as a trade mark, which means as an indicator of source or badge of origin in the course of trade, and to stop others doing the same thing with regard to a sign that is likely to cause confusion. The consequence of this is that personal use of a trade mark or use not in the context of commerce or trade to indicate source or origin is unlikely to constitute infringement.

However, Desai and Magliocca question whether home-printing, especially of a brand owner's items, could constitute commercial use for the purposes of US law—if so, then printing such items even for personal use could be infringing, even if this kind of practice does not cause confusion.[76]

Secondary liability for trade mark infringement vis-à-vis a file-sharing platform or a person uploading a file containing a registered trade mark

---

[74] Desai and Magliocca (2013–2014) 'Patents Meet Napster', 1711.
[75] Weinberg (2010) *It Will Be Awesome If They Don't Screw It Up*, p. 9.
[76] Desai and Magliocca (2013–2014) 'Patents Meet Napster', pp. 1711–1712.

to that platform may also be difficult to establish, since it must be shown that they have 'used' the trade mark in the appropriate way, that is in commerce. This may be particularly difficult to establish in platforms which permit free uploading and downloading of files, especially those which do not profit in any way from these activities. In the Internet context, keyword advertising cases internationally have generally not found the platform to be liable for secondary trade mark infringement.[77] Furthermore, jurisdictions such as Australia require that the infringing 'use' conduct takes place in Australia, which would entail that a platform or designer based, for example in the USA, may not be found liable for secondary infringement, unless there is some specific conduct directed at Australia.[78]

If an individual is printing an object which either comprises a 3D shape mark or has a 2D graphical trade mark printed on its surface, then it is highly likely that this will be personal use (unless that person tries to sell the object) and that person will not be confused as to origin as they have printed out the object in their own 3D printer.

In common law jurisdictions, trade mark claims are often brought alongside claims relating to the tort of 'passing off' (or 'misappropriation' in the USA), whereby the goodwill of a trader is appropriated in a way which causes confusion as to the origin of goods. Passing off only applies to uses which are not private as others must be misled. Where trade mark infringement cannot be made out, passing off might substitute (hence why the claims are often brought together). Bradshaw et al. consider that 3D printing may increase the scope for passing off, particularly where a trader's goodwill is comprised in the distinctive shape of goods.[79] Yet entirely private 3D printing of something that might appropriate a trader's goodwill will not constitute passing off, and if the public is not misled as to origin, then this will also not suffice to establish passing off. However, based on search engine jurisprudence, 'passing off' actions may be more successful against intermediaries than actions in secondary trade mark infringement.[80]

---

[77] Scardamaglia, A. (2014) 'Keywords, Trademarks and Search Engine Liability' in R. Konig and M. Rasch (eds.) *Society of the Query Reader: Reflections on Web Search* (Amsterdam: Institute for Network Cultures).

[78] Scardamaglia (2015) 'Flashpoints in 3D Printing and Trade Mark Law', 18.

[79] Bradshaw, Bowyer and Haufe (2010) 'The Intellectual Property Implications of Low-Cost 3D Printing', 29.

[80] Scardamaglia (2014) 'Keywords, Trademarks and Search Engine Liability'.

INTELLECTUAL PROPERTY IN PRACTICE

Much of the existing discussion around 3D printing and IP is theoretical and speculative. There is little in the way of systematic empirical studies of how IP has been interacting with 3D printing in practice. Moilanen et al. have examined how Thingiverse users share their 3D printed design files via the Thingiverse online platform, with a particular focus on licence choice, concluding that open sharing among users is less all-encompassing than Thingiverse's rhetoric would suggest.[81] This has been supplemented by a more comprehensive study of online 3D printing platforms by Mendis and Secchi, which found that 3D printing file-sharing sites are 'more expert-oriented rather than open to the masses', and that 'activity on 3D printing online platforms is [not] a mass phenomenon yet'.[82] In addition, Reeves and Mendis have conducted a study of how 3D printing is being used within industry, which will be discussed in more detail below.[83] In the following subsections, various real-life scenarios implicating 3D printing and IP are outlined and discussed.

File-Sharing Disputes

There have been various controversies involving 3D printing design files uploaded to online platforms being alleged to infringe the IP of others. In their study mentioned above, Mendis and Secchi found that there is 'evidence of intellectual property infringement, albeit on a small scale, on these online platforms' which 'highlight[s] the potential for future intellectual property issues'.[84]

Thingiverse in particular has been the recipient of various notices from IP owners relating to files uploaded to its platform which are alleged to infringe their rights. A popular means of issuing these notices has been under the takedown provisions of the US DMCA. These provisions

---

[81] Moilanen, Daly, Lobato and Allen (2015) 'Cultures of Sharing in 3D Printing: What Can We Learn from the Licence Choices of Thingiverse Users?'.

[82] D. Mendis and D. Secchi (2015) 'A Legal and Empirical Study of 3D Printing Online Platforms and an Analysis of User Behaviour' (Study I, UK Intellectual Property Office), 40–41.

[83] P. Reeves and D. Mendis (2015) 'The Current Status and Impact of 3D Printing Within the Industrial Sector: An Analysis of Six Case Studies' (Study II, UK Intellectual Property Office).

[84] Mendis and Secchi (2015) 'A Legal and Empirical Study of 3D Printing Online Platforms and an Analysis of User Behaviour, 41.

provide a 'safe harbor' to intermediaries, including online content platforms and Internet service providers, against liability for copyright infringement by their users, providing they fulfil certain requirements. These include the prompt blocking/removal of material to which these intermediaries provide access once they receive a takedown notice—that is, notification of an infringement claim from the copyright holder or their agent.

In early 2011, a Thingiverse design incorporating the famous Penrose Triangle[85]—an illusionistic 'impossible object' that is popular with 3D printing enthusiasts—received what is thought to be the first takedown notice, allegedly for reproducing another Penrose Triangle design for 3D printing that had been uploaded to rival repository Shapeways. It is unclear whether the former design infringed the copyright in the latter: the designer who alleged copyright infringement was not the original creator of the Penrose Triangle, Oscar Reutersvald, nor is the process of converting the Penrose Triangle image to a 3D printing file a clear infringement of any copyright that might subsist in the initial idea. Furthermore, it was not clear what the copyright assertion was in: the structure itself, the design file, or the image of the Penrose Triangle. Both the design file itself and the physical object that it produces may be protected by copyright, but the independent creation of an object using a different file, which was the case here, is probably not a copyright infringement, since copyright protects the expression of an idea, rather than the idea itself.

The legal status of puzzle-like 3D objects such as the Penrose Triangle, which are based on ideas, is rather complex. As mentioned above, there is some protection of physical objects—US copyright law applies to 'pictorial, graphic and sculptural works', including 'technical drawings, diagrams and models'—but 'useful articles' are excluded from copyright protection. The original 3D Penrose Triangle design, which is unlikely to be considered a useful object, was based on the 2D design from the 1930s, which is now in the public domain. 3D designs can also be conceptualised as independent interpretations of the public domain 2D original, rather than copies of the first 3D design, and so are probably not infringements, assuming copyright actually subsists in the original 3D design in the first place.[86] In any event, Thingiverse complied with the takedown request

[85] J. Wong (2011) *Penrose Triangle Illusion*, MakerBot Thingiverse, http://www.thingiverse.com/thing:6474, accessed 11 September 2015.

[86] B. Rideout (2011) 'Printing the Impossible Triangle: The Copyright Implications of Three-Dimensional Printing' *Journal of Business Entrepreneurship and the Law*, 5(1), 160–177, 170.

by removing the controversial design, 'but eventually public outcry convinced Schwanitz to dedicate his design to the public domain and retract the takedown request'.[87]

Later in 2011, the Penrose Triangle incident was followed by a more high-profile takedown notice issued by Games Workshop (the owner of Warhammer) concerning a Warhammer-style figurine designed by a Thingiverse user.[88] Thingiverse complied with the notice and removed the designs for the figurines. Again, it is unclear whether these files actually infringed copyright since the figurines seemed to be a kind of 'fan art' inspired by Warhammer, rather than a direct copy of official Warhammer figures. Indeed, the figurines 'may well have been better characterized as non-infringing original works inspired by Warhammer pieces than as infringing copies or derivative works of Warhammer pieces'.[89] The designer's 'main mistake' may have been to associate his designs with Warhammer, thus drawing attention from Games Workshop, yet in terms of legal liability, at most this may be a trade mark or trade dress infringement, or misappropriation (passing off)—which are not covered by DMCA takedown notices.[90]

A third takedown notice controversy concerning 3D printing objects occurred in January 2013, when a Tintin rocket design was also allegedly removed from Thingiverse following a DMCA notice.[91] Here, the design was based on drawings by Tintin creator Hergé in two published works, *Destination Moon* and *Explorers on the Moon*, which would still be under copyright protection according to the 'life plus 70' terms contained in the DMCA, as Hergé died in 1983. Their reproduction in the form of this design would be the strongest candidate of the examples listed here to be an actual copyright infringement, as well as possibly being another case of trade dress infringement or misappropriation due to wrongful association with the original creator.

---

[87] Weinberg (2013) *What's The Deal With Copyright and 3D Printing*, p.6.

[88] C. Thompson (2012) '3D printing's forthcoming legal morass', Wired, http://www.wired.co.uk/news/archive/2012-05/31/3d-printing-copyright, accessed 11 September 2015.

[89] Brean (2013) 'Asserting Patents to Combat Infringement via 3D Printing', 812.

[90] J. Andersen and J. Howells (2014) 'The Intellectual Property Rights Implications of Consumer 3D Printing' (Thesis, Aarhus University Department of Business Administration School of Business and Social Sciences), p.32.

[91] A. Kahler (2013) 'I got a DMCA takedown notice from Makerbot/thingiverse for this', Google+, https://plus.google.com/112825668630459893851/posts/e7sZ8Gw6umx, accessed 11 September 2015.

Yet, regardless of whether there have actually been copyright infringements in practice, the DMCA takedown mechanism is appealing to those who wish to prevent the further dissemination of designs such as those detailed above. Intermediary platforms like Thingiverse are responsive to DMCA requests, lest they lose their 'safe harbor' against potential secondary liability for copyright infringement,[92] meaning that such notices become an effective means to enforce takedown regardless of the validity of the infringement claim. There is no equivalent to this process for other IPRs, such as patents and trade marks, which thus provides an incentive for claims to be framed in copyright terms even if in practice copyright may not even subsist in the relevant file or object. In other words, rightsholders are increasingly turning to the takedown notice model—which can be easily scaled, automated, and outsourced to third parties as well—as a key weapon in their in their IP protection arsenal, even when the legal foundations for such notices are questionable. As Seng notes, the takedown process is currently 'the mainstay of content providers for managing online infringement because it is fast, cheap and efficient', partly due to it 'bypass[ing] judicial oversight over copyright disputes'.[93] The end result is a 'chilling effect' whereby even material that may not infringe copyright is still taken offline on receipt of such a takedown notice.

The use of, and reaction to, DMCA takedown notices also evidences the *de facto* application of US law over Thingiverse and its users. While the site's Terms of Use assert that it operates under New York State law, Thingiverse users are not all geographically based in that jurisdiction—yet US law prevails when it comes to takedown and removal disputes.

Despite the 'chilling effects' of takedown notices producing false positives, the more recent 'Left Shark' controversy shows some resistance from individuals to overly broad requests to take down 3D printing design files. The origin of this story was well-known pop star Katy Perry's performance at the 2015 US Super Bowl, during which she had two dancers dressed in shark costumes, one of whose awkward dance moves captured the imagination of social media users and became an Internet meme. A designer named Fernando Sosa created a 3D printing design of 'Left Shark' and uploaded the file to Thingiverse and Shapeways, but shortly afterwards received letters from Katy Perry's lawyers ordering him to remove the files

---

[92] Brean (2013) 'Asserting Patents to Combat Infringement via 3D Printing'.

[93] D. Seng (2014) 'The State of the Discordant Union: An Empirical Analysis of DMCA Takedown Notices', *Virginia Journal of Law and Technology*, 18(3), 370–473, 376.

from both platforms, as well as a takedown notice to Shapeways which removed the item from sale (although at the time of writing Shapeways had reversed its decision and the item is now back on sale).[94] Perry's lawyers claimed that the design files and shark sculpture infringed her copyright. However, Sosa decided not to comply with the demand, and enlisted help from a NYU Law School professor, who responded on Sosa's behalf, challenging Perry's lawyers' claims that copyright subsisted in the costume (when US jurisprudence has found costumes generally not to be copyrightable), and even if copyright did subsist, it certainly was not clear that Perry would be its owner.[95] In response to this, Perry's lawyers appear to be attempting to register trade marks concerning the Left Shark name and appearance (ironically using an image of Sosa's shark design), in a move that Sosa has denounced as 'trademark trolling'.[96]

Corporate Appropriation?

Another interesting phenomenon at the interaction of 3D printing and IP in practice is the attempt by various 3D printing corporations to appropriate users' creations, in sharp contrast to the scenario with Internet piracy, where alleged infringement were being carried out by individuals of large corporate copyright holders' exclusive rights.

Moilanen et al. have explored the relationship between MakerBot, parent of Thingiverse, and its users regarding the sharing of IP between the two groups.[97] While MakerBot uses rhetoric of sharing and openness to encourage users to make their creations freely available, it has been less willing to share its own IP with these users, by: moving from open-hardware principles to closed designs for its Replicator machines; altering Thingiverse's Terms and Conditions in a way which gave the company

---

[94] F. Sosa (2015a) 'Left Shark', Thingiverse, http://www.thingiverse.com/thing:667127, accessed 11 September 2015.

[95] F. Sosa (2015b) 'Political sculptor retains legal representation and responds to Katy Perry's Law Firm', http://politicalsculptor.blogspot.com.au/2015/02/politicalsculptor-retains-legal.html, accessed 11 September 2015.

[96] F. Sosa (2015c) 'Prior Art claim', http://politicalsculptor.blogspot.com.au/2015/02/prior-art-claim.html, accessed 11 September 2015; F. Sosa (2015d), 'Katy Perry Law Firm responds and so does Political Sculptor', http://politicalsculptor.blogspot.com.au/2015/02/katy-perry-law-firm-responds-and-so.html, accessed 11 September 2015.

[97] Moilanen, Daly, Lobato and Allen (2015) 'Cultures of Sharing in 3D Printing: What Can We Learn from the Licence Choices of Thingiverse Users?'.

more freedom to use users' uploaded designs for its own commercial purposes; and MakerBot's filing of 3D printing-related patent applications.

Some of these patent applications, relating to technical improvements in 3D printing hardware, have proved particularly controversial among the 3D printing maker community since it seems that similar inventions had previously been published online using open licences by community members. However, MakerBot argued that these patent applications had been filed prior to these users posting their creations online. Of course, timing here is of the utmost importance since for these patent applications to be successful, the proposed invention must exhibit the quality of novelty, which will not be the case if these user-generated ideas were already in existence at the time of the patent application.

3D Printing, Intellectual Property, and Industry

More divergence between the interaction of 3D printing and IP in practice compared to the interaction between the Internet and IP appears to be the willingness of incumbent industries to embrace 3D printing, which may in practice limit its 'disruptive' tendencies.

As mentioned above, Reeves and Mendis have conducted empirical research into industrial uses of 3D printing and IP. While they found that there was more potential for 3D printing to be used in various markets than currently, they did give examples of existing companies in the automotive sector, domestic appliances sector, customised products sector, and computer games sector already using 3D printing at least in an experimental way. Some explanation of why 3D printing was not more widespread included the fact that there were better processes available to manufacture necessary items or it was not economically viable to manufacture them using 3D printing.

Furthermore, for most industries surveyed, there was not a strong threat from consumer or prosumer 3D printing. In particular, the 'low maturity' of consumer-oriented 3D printers entails that 'the manufacturing of replacement parts for domestic appliances at home will continue to be carried out primarily by DIY and 3D printing enthusiasts', 'a niche community, which posed no perceived threat'.[98]

---

[98] Reeves and Mendis (2015) 'The Current Status and Impact of 3D Printing Within the Industrial Sector', pp. 27–28.

Yet certain company representatives, academics, and thought leaders interviewed for the research did consider that mass customisation and product personalisation by consumers would be disruptive to many kinds of businesses and would force them to change their business models, but there would still be barriers to adoption such as the skill that 3D printing requires, which would only increase as the complexity of these machines also increases.[99] For the moment though, this kind of mass customisation is likely to remain only a small proportion of the market due to the high cost of good-quality 3D printing, the low capabilities of consumer-oriented 3D printing, and the lack of appropriate design tools for consumers.[100]

Finally, Reeves and Mendis found that industry representatives were already taking steps to protect their 3D printing-related IP. Some designers they interviewed were already using encryption to protect their work—not only to protect their own IP from being infringed by others but also to conceal the fact their own designs might be infringing the IP of others.[101] Although not mentioned by their interviewees, other methods of protecting 3D printing-related IP include the use of TPMs on 3D printers themselves and design files to ensure only 'approved' files are printed, as well as streaming services such as Authentise whereby consumers can stream designs but do not have full access to the design file.[102]

Reeves and Mendis conclude their report with the view that 3D printing is unlikely to pose significant challenges to the UK's current IP framework within the next ten years due to the limitations of the technology (especially the consumer-focused machines).[103]

As an advanced developed economy, these findings for the UK are likely to be applicable to other similar jurisdictions such as the others under consideration in this book.

---

[99] Reeves and Mendis (2015) 'The Current Status and Impact of 3D Printing Within the Industrial Sector', p. 41.

[100] Reeves and Mendis (2015) 'The Current Status and Impact of 3D Printing Within the Industrial Sector', p. 42.

[101] Reeves and Mendis (2015) 'The Current Status and Impact of 3D Printing Within the Industrial Sector', p. 65.

[102] 3ders (2014) 'Authentise launches streaming service for 3D print files', http:// www.3ders.org/articles/20140404-authentise-launches-streaming-service-for-3d-print-files.html, accessed 11 September 2015.

[103] Reeves and Mendis (2015) 'The Current Status and Impact of 3D Printing Within the Industrial Sector', p. 67.

The Risks of Overenforcement

Thus, the preceding discussion demonstrates that 3D printing may not be such a significant threat to IP owners as has been argued by some. However, the overprotection and overenforcement of IPRs may also be observed, particularly through the use of DMCA takedown notices in situations where there may not be a copyright infringement. Furthermore, the interest in using DRM/TPMs to protect IP can also have problematic consequences in terms of overreach.

TPMs are designed to restrict technically what a user can do with a design file or machine, with the intention of protecting IPRs and preventing their infringement by limiting a user's ability to copy, lend, or modify files and/or use a machine in particular ways. However, in practice, the use and application of TPMs often goes beyond this protection of IP by restricting the interoperability of file formats and devices or preventing legal uses of works and items protected by IP, such as users availing themselves of one of the exceptions to protection such as fair use/fair dealing.[104]

The law on this point was harmonised by the 1996 WIPO Copyright Treaty, by which states must provide adequate legal protection and effective legal remedies against the circumvention of effective technological measures that are used by authors to protect their work. This has been transposed into the domestic laws of signatory states, which include the USA, EU, and Australia. While these transpositions usually provide exceptions to the illegality of circumventing TPMs, it is often a grey area as to whether breaking such TPMs to ensure interoperability is legal. Although hackers have worked out technical measures to get around, remove, or 'break' most forms of TPMs, these will be 'legally unattractive option[s]' for many users.[105]

The consequences of a company using TPMs on its content and/or devices can encompass consumer lock-in and a lack of interoperability with other vendors' products and services, thereby restricting consumers' choices and resisting competition from other sources.[106] The result of this can be that a particular entity builds up a dominant position in a particular market.

---

[104] See: N. Zingales (2012) 'Digital Copyright, "Fair Access" and the Problem of DRM Misuse' (Boston College Intellectual Property & Technology Forum).

[105] D. Mac Sithigh (2013) 'App law within: rights and regulation in the smartphone age' *International Journal of Law and Information Technology*, 21(2), 154–186, 170.

[106] Zittrain (2008) *The Future of the Internet and How to Stop It*, p. 177.

TPMs and the law surrounding them have been a source of controversy for some time, predating the entry of 3D printing into the mainstream. However, the interest from the 3D printing industry in using such techniques to stave off potential IP infringement should be viewed critically, given the ways in which they restrict what users can do with the design files and machines, including restricting perfectly legal uses.[107] Given the consolidation of 3D printing markets around the Big Two corporate groups of Stratasys and 3D Systems, the use of DRMs/TPMs by such entities should give cause for concern around the competitivity of 3D printing markets and the prospects of consumers and prosumers being 'locked-in' to a particular ecosystem. Furthermore, competition laws around abuse of dominance will only, in certain, exceptional, circumstances view dominant players preventing interoperability with the products and services of others as an illegal abuse.[108]

Conclusion

This chapter has explored the complex interaction between 3D printing and IP. While 3D printing has been viewed as a 'disruptive technology' for IP law, and in theory does have something of a disruptive character, in practice, the picture is rather different. The DMCA takedown notice mechanism has been used vis-à-vis file-sharing sites, and has resulted in them removing certain files uploaded by users alleged to infringe IP, even if in practice there may not be a copyright infringement. TPMs, which can also constitute an IP overreach, are also being considered in 3D printing contexts to limit what users can do with design files and their machines. Moreover, unlike the rhetoric that 3D printing would be disruptive to many incumbent industries, companies from these sectors are investigating how to use 3D printing themselves, and many do not see consumer-oriented 3D printing as a major threat to their operations and IP at the current time. To what extent this picture will change with advances in 3D printing technology and the availability of better-quality machines at lower costs remains to be seen. Yet, for the moment, the potential for 3D printing to disrupt IP laws is not really being realised in practice.

[107] See: M. Weinberg (2015) '3D Printed Copyright Creep', Techdirt, https://www.techdirt.com/articles/20150427/10532430809/3d-printed-copyright-creep.shtml accessed 11 September 2015.

[108] I. Graef, J. Verschakelen and P. Valcke (2013) 'Putting the Right to Data Portability into a Competition Law Perspective' *Law: The Journal of the Higher School of Economics Annual Review*, 53–63.

CHAPTER 3

Replicating Ruin: Printing Dangerous Objects

Abstract This chapter will consider the possibilities of using 3D printers to print dangerous or otherwise undesirable objects. Such objects lie along a spectrum of severity, from the printing of firearms and other weapons to the printing of objects which may pose mild product liability or health and safety concerns. The decentralised nature of production via 3D printing thus raises novel problems in this area, since in the previous era of mass production there have been certain 'gatekeepers' which regulate the production and circulation of these productions and accordingly can themselves be regulated, such that the products produced and transited adhere to certain standards, and that objects such as weapons are subject to strict controls regarding sale, possession and use.

At one end of this spectrum, this chapter will study the most notorious use of 3D printing to date, namely the development of the 'Liberator' 3D printed gun by the company Defense Distributed, inspired by libertarian ideology to facilitate citizens' circumvention of legislation controlling arms, and thus government control. Yet 3D printing weaponry and other items have also been of interest to nation-states: for instance, the American and Australian armies have conducted research into 3D printing's applications in the military and challenges for the military. Here, the application of relevant laws controlling arms—both internally and for export—to 3D printing militias will be considered, along with the idiosyncratic situation of the

© The Editor(s) (if applicable) and The Author(s) 2016
A. Daly, *Socio-Legal Aspects of the 3D Printing Revolution*,
DOI 10.1057/978-1-137-51556-8_3

49

USA, whereby the bearing of arms is enshrined as a constitutional right. The Liberator phenomenon has also sparked legislative responses in various places, so an assessment will be made of how successfully (or not) these law enforcement agencies have been able to contain 3D printed weaponry.

Objects at the other, less destructive, end of this spectrum will also be considered, looking at the issues for product liability and health and safety standards they may pose. One major application of 3D printing so far has been in the health field, including the printing of customised prosthetics, medical implements, and biological matter. The 3D printing of prosthetics and implements in particular is illustrative of the dilemma between lowering costs and increasing availability for such medical products particularly among poor communities throughout the world, while sidestepping the regulation which ensures such products are of an adequate standard. The implications of these developments for the effective enforcement of existing law and regulation in this area will be considered.

THE LIBERATOR AND GUN CONTROL

The issue of 3D printed guns came to attention in 2013 when US-based company Defense Distributed developed blueprints for a gun, the Liberator, which could be created using a 3D printer. To date, this is the best known and most notorious example of the possibility of creating dangerous objects using 3D printers.

Defense Distributed was founded by former Texan law student Cody Wilson in 2012, who acts as the organisation's spokesperson. The company made its initial appearance with a website and Indiegogo crowdfunding campaign with the aim of raising US$20,000 in order to design and release blueprints for a plastic gun which could be created with an entry-level 3D printer.[1] However, a few weeks later, Indiegogo suspended Defense Distributed's campaign for violating the site's terms of service, refunding the money which was pledged.[2] Donations to Defense Distributed's activities could still be made using Bitcoin via its website (which remains the case at the time of writing), and by the end of 2012, the organisation had raised

---

[1] A. Greenberg (2012) '"Wiki Weapon Project" Aims to Create A Gun Anyone Can 3D-Print At Home', Forbes, http://www.forbes.com/sites/andygreenberg/2012/08/23/wiki-weapon-project-aims-to-create-a-gun-anyone-can-3d-print-at-home/, accessed 11 September 2015.

[2] F. Martinez (2012) 'Indiegogo shuts down campaign to develop the world's first printable gun', Daily Dot, https://www.dailydot.com/news/indiegogo-3d-printed-gun-campaign/ accessed 11 September 2015.

enough money to fund its initiative. Part of this money was spent on renting a Stratasys 3D printer, which Stratasys subsequently recalled due to its concerns that Defense Distributed was using the printer for 'illegal purposes'.[3]

Despite these setbacks, in January 2013, Defense Distributed publicly released files for AR-15 standard capacity magazine, which could be printed on a 3D printer, and in March of that year it released files for a durable printed AR-15 receiver. It also created a gun file repository called DEFCAD to distribute these files, which was important since in late 2012 Thingiverse had removed files from its site, which comprised blueprints for gun parts, and changed its terms of service to exclude content which 'promotes illegal activities or contributes to the creation of weapons'.[4]

In May 2013, two months after Cody Wilson was granted a federal firearms license to manufacture and deal in firearms,[5] Defense Distributed released the files for its 'Liberator' pistol, the world's first entirely 3D printed (or printable) gun, only requiring a commonplace metal nail to be inserted as its firing pin.[6] The Liberator's design files, uploaded to Defense Distributed's website, were downloaded more than 100,000 times in the space of two days alone.[7]

The Liberator was test-fired shortly after its release,[8] demonstrating that it could be printed out on a consumer-oriented 3D printer and could also function as a firearm. However, acquiring a gun in this way remains onerous, compared to alternatives—as Wilson himself has acknowledged:

> It's already possible, if you want, to just go buy some pipes and put a gun together. This is another thing that I think has been conflated. People thought, 'oh no this is the first time now people can expediently make guns.'

[3] K. Streams (2012) '3D printed gun project halts after Stratasys confiscates rented printer', The Verge, http://www.theverge.com/2012/10/1/3439496/wiki-weapon-project-defense-distributed-stratasys, accessed 11 September 2015.

[4] T. Maly (2012) 'Thingiverse Removes (Most) Printable Gun Parts', Wired, http://www.wired.com/2012/12/thingiverse-removes-gun-parts/, accessed 11 September 2015.

[5] C. Farviar (2013) '3D-printed gun maker now has federal firearms license to manufacture, deal guns', Arstechnica, http://arstechnica.com/tech-policy/2013/03/3d-printed-gunmaker-now-has-federal-firearms-license-to-manufacture-deal-guns/, accessed 11 September 2015.

[6] A. Greenberg (2013a) 'Meet The 'Liberator': Test-Firing The World's First Fully 3D-Printed Gun', Forbes, http://www.forbes.com/sites/andygreenberg/2013/05/05/meet-the-liberator-test-firing-the-worlds-first-fully-3d-printed-gun/, accessed 11 September 2015.

[7] A. Greenberg (2013b) '3D-Printed Gun's Blueprints Downloaded 100,000 Times In Two Days (With Some Help From Kim Dotcom)', Forbes, http://www.forbes.com/sites/andygreenberg/2013/05/08/3d-printed-guns-blueprints-downloaded-100000-times-in-two-days-with-some-help-from-kim-dotcom/, accessed 11 September 2015.

[8] See: Greenberg (2013a) 'Meet The 'Liberator'.

No; in fact, this is a very inexpedient way of making a gun, and kind of ridiculous. But it's trying to be demonstrative and predictive of the future.[9]

Legal or illegal alternatives of acquiring a 'manufactured' gun are likely to be more expedient to users than 3D printing one, but '[t]he likelihood is that better 3D printers will make better guns, perhaps even approaching the effectiveness, reliability, and safety of traditionally manufactured weapons'.[10] Yet, there have already been instances of people using 3D printers to make guns[11] and getting arrested and imprisoned for doing so in certain places.[12] 3D printers have also been used to make other 'undesirable' equipment used in criminal enterprises, such as ATM skimming devices, which were found in a Europol raid.[13] Furthermore, the possibilities afforded by 3D printed weapons have piqued the interest of white supremacists groups.[14]

Reaction of US Law Enforcement

All of these steps leading up to the Liberator's birth had been heavily covered in global media, with Defense Distributed's members, particularly Cody Wilson, freely giving interviews to assembled journalists. There was nothing particularly clandestine about their activities, and they quickly attracted the attention of law enforcement agencies, particularly in the USA where Defense Distributed is based.

---

[9] C. Wilson (2013) quoted in S. Paikin '3D Printing: A Killer App', The Agenda, https://www.youtube.com/watch?v=eN_cVRjIrwg, accessed 11 September 2015.

[10] I. Record, g. coons, D. Southwick and M. Ratto (2015) 'Regulating the Liberator: Prospects for the Regulation of 3D Printing' *Journal of Peer Production* Issue #6 Disruption and the Law. The authors, based in Canada, built a non-functioning Liberator using their 3D printing lab, and based on their experience remarked that 'the act of making a Liberator, remained, for the moment, impracticable for most people, for lack of access to equipment and expertise'.

[11] See: B. Ashcraft (2014) 'Japanese Man Arrested for Having Guns Made with a 3D Printer', Kotaku, http://kotaku.com/japanese-man-arrested-for-having-guns-made-with-a-3d-pr-1573358490, accessed 11 September 2015.

[12] B. Krassenstein (2014) 'Two Year Sentence Handed Down to Yoshitomo Imura in Japanese 3D Printed Gun Case', 3DPrint.com, http://3dprint.com/20019/sentence-imura-3d-printed-gun/, accessed 11 September 2015.

[13] EUROPOL (2014) '31 Arrests In Operation Against Bulgarian Organised Crime Network', https://www.europol.europa.eu/content/31-arrests-operation-against-bulgarian-organised-crime-network, accessed 11 September 2015.

[14] R. Fordyce (2015) 'Manufacturing Imaginaries: Neo-Nazis, Men's Rights Activists and 3D Printing' *Journal of Peer Production*, Issue #6 Disruption and the Law.

While in most jurisdictions the manufacture, distribution, and sale of firearms are heavily restricted, the USA has taken a different approach, as codified
in the Second Amendment to the American Constitution, which provides:

> A well regulated militia being necessary to the security of a free state, the
> right of the people to keep and bear arms shall not be infringed.

The US Supreme Court in *Heller* has interpreted this provision as giving rise to 'an individual right to possess a firearm unconnected with service in a militia, and to use that arm for traditionally lawful purposes,
such as self-defense within the home'.[15] Nevertheless, while the Second
Amendment gives rise to this presumptive right to possess a firearm, gun
rights and gun control have coexisted in the USA since the time of the
Founding Fathers.[16] Indeed, the USA has not refrained from imposing a
complex system of regulating firearms:

> Anyone "engage[d] in the business" of manufacturing, importing or deal-
> ing in firearms is required to become a federal firearm licensee ("FFL").
> At its creation, a gun must possess a serial number that the manufacturer
> is required to keep on record. Once built, firearms are sold by the FFL
> manufacturer to FFL dealers such as pawnshops and retail stores. Federal
> law requires dealers to keep records on almost all firearm transactions, and
> any transfers in interstate commerce must occur between licensees. No one,
> not even an FFL, may transfer a firearm to a person who is known or rea-
> sonably believed to be an out-of-state resident, felon or fugitive from the
> law. A private individual first comes into contact with the system when he
> or she attempts to purchase a new firearm from an FFL dealer. The pro-
> spective buyer submits to the National Instant Criminal Background Check
> System ("NICS"); if the buyer is of age, and not otherwise prohibited from
> possessing a firearm, the transfer is approved and the NICS records of the
> applicant's identity are destroyed. While the FFL dealer retains paper pur-
> chase records, those records may not be digitized or compiled into a data-
> base by the Bureau of Alcohol, Tobacco, Firearms and Explosives ("ATF"),
> the agency that promulgates and enforces firearm regulation consistent with
> federal statute.[17]

---

[15] *District of Columbia v Heller* (2008) No. 07-290.
[16] See: A. Winkler (2013) *Gunfight* (New York: W. W. Norton).
[17] P. Jensen-Haxel (2012) '3D Printers, Obsolete Firearm Supply Controls, and the Right
to Build Self-Defense Weapons Under Heller' *Golden Gate University Law Review,* 42(3),
447–495, 457.

Furthermore, the National Firearms Act 1934 renders illegal a list of weapons often used by organised criminal groups, such as machine guns, providing that manufacturers, dealers, and possessors of these weapons must undergo a specific registration process. The Undetectable Firearms Act also requires that all major gun components be designed in a way that ensures they can be identified by x-ray machines and trigger metal detectors.

In other jurisdictions, without such a strong affirmation of the right to bear arms, Defense Distributed's activities would likely have breached laws regulating the manufacture and distribution of firearms. However, the Second Amendment has complicated the US legal situation when addressing the Liberator.

Defense Distributed was ordered to take down the files from its DEFCAD site by the US State Department shortly after the files were publicly released, for alleged infringement of American arms export control laws, namely the Arms Export Control Act (AECA) and the International Traffic in Arms Regulations (ITAR), since Defense Distributed did not seek prior authorisation from the US Directorate of Defense Trade Controls before posting the files online in violation of ITAR.[18] Defense Distributed complied with the order in removing the files from its DEFCAD site, although the files could still be downloaded from other locations online. While overseas and geographically outside of US jurisdiction, Mega's owner Kim Dotcom ordered his staff to remove public links to the 3D printed gun blueprints from the cyberlocker service,[19] in an act of private regulation. Yet the files could still be accessed via The Pirate Bay in November 2013, and as of late 2014, at least it seems that the blueprints were still being circulated on file-sharing sites better known for illicit and risqué materials such as pornography and cracked software.[20]

Ongoing Legal Battle

At the time of writing, Wilson and Defense Distributed had recently commenced a legal battle against the US State Department, arguing that the

---

[18] A. Greenberg (2013c) 'State Department Demands Takedown of 3D-Printable Gun Files For Possible Export Control Violations', *Forbes*, http://www.forbes.com/sites/andygreenberg/2013/05/09/state-department-demands-takedown-of-3d-printable-gun-for-possible-export-control-violation/, accessed 11 September 2015.

[19] G. Ferenstein (2013) 'Offshore 3D Printed Gun Blueprint Protector Kim Dotcom Reportedly Deleting Files', TechCrunch, http://techcrunch.com/2013/05/11/offshore-3d-printed-gun-blueprint-protector-kim-dotcom-reportedly-deleting-files/, accessed 11 September 2015.

[20] Record, coons, Southwick and Ratto (2015) 'Regulating the Liberator'.

letter the State Department sent Defense Distributed ordering it to take down the 3D printed gun blueprints from its website violated their constitutional rights, namely their First Amendment right to free speech comprised in the blueprint files, their Second Amendment right to bear arms, and their Fifth Amendment right to due process.[21] In particular, Defense Distributed argues that the pre-publication requirement to seek authorisation from law enforcement agencies before publishing files such as those for the Liberator is an unconstitutional prior restraint on their free speech.

While it may seem far-fetched to the layperson that 3D printing design files for a functioning gun could constitute free speech and be constitutionally protected, the First Amendment has been interpreted expansively by American courts, especially the Supreme Court, as will be seen in the following sections.

History of ITAR
The purpose of the ITARs is to control the import and export of defence-related items and services from the USA with the aim of safeguarding US national security and furthering US foreign policy objectives. This regulatory regime was enacted during the Cold War with the former USSR. The items and services regulated are contained in the United States Munitions List, and the regulatory scheme permits such items and services on this list to be shared with other US 'persons' (including organisations) but their provision to foreign persons without authorisation (or the use of an exemption) is prohibited, and violations can result in large fines being imposed.

The relationship between the ITAR scheme, technology, and the US Constitution has already come under examination by the courts during the 1990s 'cryptowars'. Cryptographic software and algorithms were included in the Munitions List, and so American persons wishing to export such materials had to comply with the ITAR scheme, especially the requirement for prior authorisation before export.

This inclusion of cryptographic materials in the Munitions List was challenged by David Bernstein, who at the time was a graduate student in mathematics at the University of California, Berkeley, doing research into cryptography. Bernstein was concerned that he would be violating the ITAR regime by communicating his research on cryptography through

---

[21] *Defense Distributed and Second Amendment Foundation v US Department of State* Case No 1:15-cv-372. See: A. Greenberg (2015a) '3-D Printed Gun Lawsuit Starts the War Between Arms Control and Free Speech', Wired, http://www.wired.com/2015/05/3-d-printed-gun-lawsuit-starts-war-arms-control-free-speech/, accessed 11 September 2015.

publication, putting it on a website, or including the research as content in a class where there were students in the audience who were foreign nationals. This is because 'export' is defined broadly by ITAR to include 'disclosing (including oral or visual disclosure) or transferring technical data to a foreign person, whether in the United States or abroad'.[22] Bernstein requested clarification from the Office of Defense Trade Controls (ODTC) as to whether his activities fell within the ambit of ITAR, which replied initially stating that his activities were subject to export licensing requirements, a decision which subsequently was partially reversed, such that the restrictions applied only to the source codes for Bernstein's encryption and decryption programs.

In any event, Bernstein sought a declaratory judgement against the US Department of State to prevent it from enforcing the AECA and ITAR for incompatibility with his constitutional rights, namely his First Amendment right to free speech. The court of first instance held that cryptographic computer source code is speech covered by the First Amendment.[23] In a subsequent opinion, the court also held that the Munitions List's licensing requirement for cryptographic software was an unconstitutional prior restraint of speech and that the source code of Bernstein's programs itself was protected speech.[24] The Ninth Circuit Court of Appeals also held that encryption software in its source code was expression covered by the First Amendment and was entitled to protections of the prior restraint doctrine,[25] and that the regulations preventing the source code's publication were unconstitutional.

Another, similar case was brought by Case Western Reserve University law professor Peter Junger, who was concerned about the legality of teaching encryption in a computer law class containing non-US citizens and the legality of placing encryption programs he had created on his website. At first instance, a US District Court judge held that encryption software was not sufficiently expressive to be protected by the First Amendment,[26] but on appeal the Sixth Circuit Court of Appeal reversed this decision, holding that software source code was protected by the First Amendment.[27]

---

[22] International Traffic in Arms Regulations, 22 C.F.R. § 120.17(a)(4) (1996).

[23] *Bernstein v. U.S. Dep't of State (Bernstein I)*, 922 F. Supp. 1426 (N.D. Cal. 1996), 1437.

[24] *Bernstein v. United States Dep't of State*, 945 F. Supp. 1279 (N.D. Cal.1996) at 1287.

[25] *Bernstein v United States* (1999) Case Number: 97-16686 (9th Circuit Court of Appeal), 4234.

[26] *Junger v. Daley*, 8 F. Supp. 2d 708 (N.D. Ohio 1998).

[27] *Junger v. Daley*, 209 F.3d 481 (6th Cir. 2000).

The Court of Appeal also held that the government bears a strong burden to show that national security interests justify the kinds of prior restraint on speech contained within the ITAR regime.

In the litigation, code has been considered to be 'speech' because of its verbal characteristics rather than 'conduct'—an important distinction in American First Amendment jurisprudence since 'speech' will be more strictly scrutinised by courts than conduct.[28] Conceptualising code as speech has been viewed as 'not entirely satisfying' due to the way in which software 'inseparably incorporates elements of both expression and function' and the ways in which it may be desirable to regulate code, for example, to prevent the transmission of computer viruses.[29] The strong protection given to First Amendment speech may frustrate such attempts to regulate code in this way.

However, Tien has argued that not every 'software act' is a 'speech act', and the dissemination of viruses would not constitute a speech act because it lacks any communicative intent, which he sees as at the root of speech acts and accordingly First Amendment coverage.[30] In Tien's view, the publication of source code would be a speech act, but the use of software is different from the publication of software, which would open up the possibility of regulating that use even if speech acts involving software and code may attract First Amendment protection after the *Bernstein* litigation. Post has criticised Tien's conceptualisation of First Amendment protection pertaining to speech acts given the many speech acts which do not enjoy First Amendment protection such as product warnings, and instead, in his view, 'First Amendment coverage is triggered by those forms of social interaction that realize First Amendment values' such as social interactions which constitute a 'marketplace of ideas' or that 'instantiate the value of self-government'.[31] As regards encryption, for Post the relevant distinction was 'between encryption source code that

---

[28] See: A. O. Wertheimer (1994) 'The First Amendment Distinction Between Conduct and Content: A Conceptual Framework for Understanding Fighting Words Jurisprudence' *Fordham Law Review*, 63, 793–851.

[29] T. Nguyen (1997) 'Cryptography, Export Controls, and the First Amendment in *Bernstein v. United States Department of State*' *Harvard Journal of Law & Technology*, 10(3), 667–682, 675–677.

[30] L. Tien (2000) 'Publishing Software As a Speech Act' *Berkeley Technology Law Journal* 15, 629–712, 669.

[31] R. Post (2000) 'Encryption Source Code and the First Amendment' *Berkeley Technology Law Journal*, 15, 713–724, 716.

is itself part of public dialogue and encryption source code that is meant merely to be used'.[32]

In addition to these cases on encryption software, the US Supreme Court has extended First Amendment protection to electronic communications[33] and a broad class of 'information'.[34]

First Amendment and 3D Printed Guns

Even before Defense Distributed commenced litigation against the US State Department, American First Amendment scholars had already been debating the extent to which the State Department, in ordering the Liberator files to be removed from Defense Distributed's site, was acting in compliance with Defense Distributed's First Amendment rights (and those of people wishing to access these files). The *Bernstein* litigation provides interesting precedents, particularly since it involved the same regulations as those used by the State Department to order the Liberator to be taken down.

Precisely how design files (and/or their communication via the Internet) should be characterised for First Amendment purposes will determine the level of protection (if any) they receive from government interference—whether these files constitute speech despite their functionality. Cosans has considered design files to have both expressive and functional aspects, with sufficient communicative elements to engage the First Amendment, but since in her view the files are not 'pure speech' their restriction by the government should be assessed in accordance with whether a sufficiently important government interest exists in regulating the non-speech elements ('intermediate scrutiny')—which she considers to be the government's interest in harm reduction, which in her view justifies the State Department ordering Defense Distributed to take down the gun design files.[35]

However, differing approaches to 3D printing design files have been taken by other scholars. Langvardt considers that design files themselves are not 'speech' but 'can be used in the service of speech', and he believes that Cody Wilson's circumstances do seem to involve a 'speech act' because

---

[32] Post (2000) 'Encryption Source Code and the First Amendment', 720.

[33] *Reno v American Civil Liberties Union* 521 U.S. 844 (1997).

[34] *Sorrell v IMS Health* 131 S.Ct. 2653 (2011).

[35] J. Cosans (2014) 'Between Firearm Regulation and Information Censorship: Analyzing First Amendment Concerns Facing the World's First 3-D Printed Plastic Gun' *Journal of Gender, Social Policy and Law*, 22(4), 915–946.

'[h]is intent seems to have been to demonstrate the futility of gun control against the Internet',[36] although he believes that intermediate scrutiny is the appropriate First Amendment standard to apply here.

Nevertheless, another position has been taken by Blackman, who considers the regulation of '3D CAD source files [to be] really a regulation on information' which 'must satisfy constitutional scrutiny'; since the restriction on these files is based on the content of the source code (the object the information expresses), strict scrutiny should apply to this restriction of expression.[37] What particularly is problematic for Blackman seems to be the government restriction on distributing information about how to exercise another constitutional right, namely the Second Amendment, which he views as a prior restraint on free speech. While the US government may have a countervailing interest in the form of national security to regulate the Liberator files, previous Supreme Court jurisprudence suggests that there must be a 'direct and imminent' imperilling of national security,[38] which places a high burden on the government to show this, and Blackman considers that '[a]n open-sourced CAD file of a simple pistol that is readily available all over the internet would not even come close to meeting this lofty threshold'.[39]

Second Amendment and 3D Printed Guns

Yet it is not only the First Amendment right to free expression which is implicated in the Liberator's legal saga. The scope of the Second Amendment's right to bear arms is directly at issue in Defense Distributed's case against the US government, particularly the extent to which the Second Amendment gives rise to an individual right to make and possess a firearm, and if so, what the scope of this right is.

As mentioned above, the US Supreme Court recognised an individual right to possess firearms in its decision in *Heller*. However, it is not clear what the scope of this right is—it seems to protect weapons 'in common use at the time' (such as handguns) but not 'dangerous and unusual weapons' (such as machine guns). For Jensen-Haxel, it is not clear to what

---

[36] K. Langvardt (2014) 'The Replicator and the First Amendment', *Fordham Intellectual Property Media Entertainment and Law Journal*, 25(1), 59–115, 94.

[37] J. Blackman (2014) 'The 1st Amendment, 2nd Amendment, and 3D Printed Guns' *Tennessee Law Review*, 81, 479–538, 501.

[38] *New York Times v United States*, 403 US 713, 730 (Stewart J, joined by White J, concurring).

[39] Blackman (2014) 'The 1st Amendment, 2nd Amendment, and 3D Printed Guns', 536.

extent *Heller* protects new weapons technology, and whether 3D printed guns such as the Liberator would be considered to be within 'common use'.[40] Jensen-Haxel also views *Heller* as probably also giving rise to a right to acquire firearms as well as possess them, with a historical perspective also supporting 'a general right to make one's own arms for personal use'.[41]

Blackman believes the situation under US law to be more clear-cut: he points to guidance from the Department of Justice that states it is legal to make firearms without a licence so long as they are not for sale and the maker is not prohibited from possessing firearms.[42] While he acknowledges the differing views from the jurisprudence that lower courts have taken after *Heller* regarding the right to acquire arms being part of the Second Amendment, Blackman considers that 'a constitutional right to bear arms without a complementary right to acquire (buy and sell) arms would be meaningless'.[43]

It remains to be seen how these precedents and approaches to the First and Second Amendment may be applied in the Defense Distributed litigation, and thus to what extent Americans at least are entitled to access information online about how to make a 3D printed gun.

Other Legal Responses to the Liberator

Alongside this chain of events, legislative approaches have been pursued both in the USA and elsewhere to address the Liberator. This section will detail these attempts.

USA

Some American legislators have been greatly concerned by the emergence of the 3D printed Liberator gun. Steve Israel, Democratic Representative for New York, has stated that he wishes to ban 3D printed gun components and 3D printed ammunition attachments as part of the renewal

---

[40] Jensen-Haxel (2012) '3D Printers, Obsolete Firearm Supply Controls, and the Right to Build Self-Defense Weapons Under Heller'.

[41] Jensen-Haxel (2012) '3D Printers, Obsolete Firearm Supply Controls, and the Right to Build Self-Defense Weapons Under Heller', 479.

[42] See: U.S. Department of Justice Bureau of Alcohol, Tobacco, Firearms and Explosives (2015) 'What is ATF doing in regards to people making their own firearms?', https://www.atf.gov/firearms/qa/what-atf-doing-regards-people-making-their-own-firearms, accessed 11 September 2015.

[43] Blackman (2014) 'The 1st Amendment, 2nd Amendment, and 3D Printed Guns', 495.

of the Undetectable Firearms Act, since these parts would not be easily identified using methods such as metal detectors.[44] In the end, the Act was extended for another ten years from December 2013 but no further prohibitions were included in the extension beyond those in the original version of the law.[45] Israel wanted to see requirements introduced that certain major components of plastic firearms are non-removable and made of materials that are detectable, that is, made of metal.[46]

While Israel failed to pass this legislation in 2013, at the time of writing it has been reported that he is planning to reintroduce the Undetectable Firearms Modernization Act.[47] Although 3D printed weapons such as the Liberator seemed the overt target of Israel's initial attempt at legislation, his current proposals appear technology neutral inasmuch as they do not explicitly single out 3D printing as a manufacturing technique, and instead focus on the characteristics of the weapon itself.[48]

In November 2013, the American city of Philadelphia passed a law which prohibited those without a federal firearms licence from using 3D printers to make guns.[49]

Australia

3D printing and the Liberator gun have strongly captured the Australian national imagination. In 2013, following the media storm around the Liberator, the New South Wales police force bought a 3D printer, down-

[44] See: C. Doctorow (2012) 'Congressman calls for ban on 3D printed guns', Boing Boing, http://boingboing.net/2012/12/09/congressman-calls-for-ban-on-3.html, accessed 14 June 2015.

[45] D. Roberts (2013) '3D-printed guns prompt US House to renew prohibition on plastic firearms', The Guardian, http://www.theguardian.com/world/2013/dec/04/3d-guns-house-renew-prohibition-plastic-firearms, accessed 12 September 2015.

[46] S. Israel (2013) 'Rep. Israel Introduces Bipartisan Undetectable Firearms Modernization Act to Protect Americans from Threat of Plastic Guns', press release, http://israel.house.gov/media-center/press-releases/rep-israel-introduces-bipartisan-undetectable-firearms-modernization-act, accessed 14 September 2015.

[47] A. Greenberg (2015b) 'Bill to Ban Undetectable 3D Printed Guns Is Coming Back', Wired, http://www.wired.com/2015/04/bill-ban-undetectable-3-d-printed-guns-coming-back/accessed 12 September 2015.

[48] M. Molich-Hou (2015) 'Rep. Steve Israel Renews Fight for Undetectable Gun Control', 3D Printing Industry, http://3dprintingindustry.com/2015/06/10/rep-steve-israel-renews-fight-for-undetectable-gun-control/, accessed 12 September 2015.

[49] I. Volsky (2013) 'Philadelphia Becomes First City To Ban 3D Guns', Think Progress, http://thinkprogress.org/justice/2013/11/23/2987911/philadelphia-city-ban-guns/, accessed 12 September 2015.

loaded the blueprints, and built their own 3D printed weapon in 27 hours.[50] In testing the printed gun, the weapon encountered a 'catastrophic failure' when it was fired, with the barrel exploding, demonstrating the risks posed to the bearer of the weapon as well as its target.[51] Yet, what appeared to be 3D printed gun parts were found among other illicit items in a police raid on a Gold Coast property in early 2015.[52]

The state of Queensland was the site of an attempt to outlaw 3D printed guns in 2014.[53] The Palmer United Party representatives in Queensland's Parliament introduced a Private Members' Bill which would have made it a punishable offence to make, buy, or possess 3D printed firearms and design files for such weapons without prior authorisation.[54] A Queensland parliamentary committee advised against passing the bill in November 2014, the bill lapsed when the parliament was dissolved prior to state elections and the new government did not adopt the bill, pointing to the existing legislation in the state which already addressed the unlawful manufacture of weapons.[55]

3D printed guns have also been considered by the federal parliament in Australia, in the context of a Senate Inquiry into gun-related violence during 2014 and 2015. Unlike the USA, in Australia there is no constitutional right to bear arms (nor even an explicit constitutional right to free expression). Since various fatal mass shootings in the 1980s and 1990s, Australia introduced more restrictive weapons regulation, which requires those wishing to possess a firearm to obtain a prior licence and demonstrate a 'genuine reason for owning, possessing or using a firearm' in addition to other requirements. Firearm sales must also only be conducted by or through licensed dealers, and each firearm must be registered.

[50] I. Gridneff (2013) '3D-printed gun 'will kill', police warn', Sydney Morning Herald, http://www.smh.com.au/digital-life/digital-life-news/3dprinted-gun-will-kill-police-warn-20130524-2k59g.html, accessed 12 September 2015.

[51] R. Pearce (2013) 'NSW Police issues warning on 3D printed guns', Computer World, http://www.computerworld.com.au/article/462774/nsw_police_issues_warning_3d_printed_guns/, accessed 12 September 2015.

[52] R. Varley and M. Eaton (2015) '3D printing: Suspected plastic gun parts found in raid on Gold Coast property', ABC, http://www.abc.net.au/news/2015-02-10/3d-printing-police-suspect-plastic-parts-belong-to-homemade-gun/6083938, accessed 12 September 2015.

[53] E. Worthington (2014) '3D printed guns: PUP introduces Queensland bill to regulate digitally generated firearms', ABC, http://www.abc.net.au/news/2014-05-23/3-d-printed-guns-palmer-party-introduces-qld-bill-3d-firearms/5472566, accessed 12 September 2015.

[54] Weapons (Digital 3D and Printed Firearms) Amendment Bill 2014 (QLD).

[55] P. Cowan (2015) 'Qld Govt knocks back 3D-printed guns bill', IT News, http://www.itnews.com.au/News/403827,qld-govt-knocks-back-3d-printed-guns-bill.aspx, accessed 12 September 2015.

The Committee issued a report in April 2015—or in fact, two reports: one from the Chair of the Committee and one other Senator ('the Chair's report'), and the second report from four other members of the Committee ('a majority of Senators').

The Chair's report recommended that:

- The Australian state and territory governments investigate the requirement for uniform regulations in all jurisdictions covering the manufacture of 3D printed firearms and firearm parts;
- The Australian state and territory governments continue to monitor the risks posed by the manufacturing of firearms using 3D printers and consider further regulatory measures if necessary.[56]

The former recommendation can be contextualised by the fact that the existing regulatory framework for firearms in Australia comprises a variety of laws in different states and territories which are not always consistent with each other. However, the Chair's report noted that '[i]t seems that current laws pertaining to firearms would apply equally to 3D printed firearms and firearm parts'.[57] Aside from harmonising these laws, it is unclear why the Chair's report recommended the introduction of legislation specifically regulating 3D printed firearms and firearm parts. Yet the Chair's report also acknowledged the danger of overregulating 3D printing, and recommended against 'banning the individual use of 3D printers or introducing a character test for ownership'.[58] Interestingly, the dissenting report from a majority of the Senators on the Committee considered that 'Commonwealth, State and Territory laws relating to the import and manufacture of firearms or firearm parts, including by 3D printers, was [sic] sufficient to enable prosecution of any offence',[59] and accordingly they believed that there was no need for new regulations to be introduced

---

[56] Australian Senate Legal and Constitutional Affairs References Committee (2015) *Ability of Australian law enforcement authorities to eliminate gun-related violence in the community*, p. xii.

[57] Australian Senate Legal and Constitutional Affairs References Committee (2015) *Ability of Australian law enforcement authorities to eliminate gun-related violence in the community*, p. 93.

[58] Australian Senate Legal and Constitutional Affairs References Committee (2015) *Ability of Australian law enforcement authorities to eliminate gun-related violence in the community*, p. 93.

[59] Australian Senate Legal and Constitutional Affairs References Committee (2015) *Ability of Australian law enforcement authorities to eliminate gun-related violence in the community*, p. 144.

to cover the manufacture of 3D printed firearms and firearm parts at this point in time.[60]

Subsequent to this report, at the time of writing no concrete steps have been taken to alter existing Australian legislation to take account of the phenomenon of 3D printed guns.

Post-Control?

As mentioned in Chap. 1, the creation of the Liberator and the hype around it have inspired some to believe that 3D printing is driving a 'post-control' society whereby the nation-state is no longer able to enforce its laws. In practice, there have been few occasions on which 3D printed guns have been found or used. It would seem that traditional gun-making and gun-sourcing methods, whether licit or illicit, are still to be preferred. Reasons for this would seem to be the limited functionality of 3D printers, especially those marketed at consumers.

Indeed, a team of academic researchers at the University of Toronto constructed a non-functioning version of the Liberator, remarking that:

> What struck us first was the sheer size of the infrastructure needed to manufacture the object. It took two PhD students and a post-doc the better part of three days to make the object. In making it they had engaged with tens of thousands of dollars worth of equipment, each piece with its own supporting infrastructure. To us, this served as a clear demonstration that the act of making a Liberator remained, for the moment, impracticable for most people, for lack of access to equipment and expertise. Finally, everyone who worked on the project at one point or another commented on how rudimentary the Liberator was, in terms of both functionality and feel. None of us felt comfortable with the idea of test-firing the object by hand.[61]

This demonstrates the impracticality of making Liberators for most people, and goes some way to dispelling the idea that since now everyone has the means to make firearms, the power of the nation-state is greatly diminished. Moreover, as also mentioned in Chap. 1, the apparatus of nation-states is also engaging with 3D printing for their own defence

---

[60] Australian Senate Legal and Constitutional Affairs References Committee (2015) *Ability of Australian law enforcement authorities to eliminate gun-related violence in the community*, p. 149.

[61] Record, coons, Southwick and Ratto (2015) 'Regulating the Liberator: Prospects for the Regulation of 3D Printing'.

efforts, as both a threat and an opportunity for those efforts. Thus, 3D printing presents new possibilities for both those within and those opposing nation-states, but at the present time it would seem that a post-control situation in which the nation-state and its weapons laws are redundant is not what is currently being experienced.

CONSUMER SAFETY

Although less sensational for the popular imagination than the danger presented by the Liberator, 3D printing objects also present safety problems for consumers, in particular how the health and safety standards to which conventionally manufactured items have typically been subjected can be upheld. Ironically, the Liberator itself presents some of these consumer safety problems, inasmuch as some attempts to make the gun on a 3D printer have resulted in a defective version being produced which may be as dangerous for the person firing the gun as the target of the shot, as the New South Wales police have demonstrated. Product safety concerns may be particularly pronounced for items produced with cheaper lower-quality 3D printers. Yet, 3D printing may have some benefits from a product safety perspective. Small runs of 3D printed products 'reduce the reach of any product defects' and make it easier to recall such defective products.[62]

In this section, two areas of law which aim to ensure consumer safety will be examined: product liability and then medicines and medical device regulation. The former area has attracted some academic commentary although it seems that no cases have been brought regarding defective 3D printed products at the time of writing. The latter area is also significant given the prominent medical applications of 3D printing so far.

Product Liability Law

Product liability laws can be conceptualised as aiming to ensure the safety of consumers, on the one hand, and, on the other hand, ensuring a chain of accountability via manufacturers, distributors, suppliers, and retailers for the consumer harm caused by defective products, and providing these entities with incentives to ensure they take proper care in making and distributing such products. Current product liability laws are usually pre-

---

[62] G. Greatorex (2015) '3D Printing and Consumer Product Safety', Product Safety Solutions White Paper, p. 13.

mised on the idea that the party being held liable is that best able to shoulder the burden of such liability, especially vis-à-vis the individual consumer since those entities being held liable are likely to be large(r) commercial operations which are well placed to seek insurance to cover their legal liabilities. 3D printing by prosumers provides a challenge to these laws and this conceptual basis underlying them, as will be detailed.

Some scholars, particularly from the USA, have begun to discuss the relationship between 3D printing and liability for defective products.[63] The relevant US legislation on this issue is the Third Restatement of Torts, which provides that a product is 'defective' if it has a manufacturing defect, a design defect or if it is accompanied by inadequate instructions or warnings.[64] Other main actions in product liability are for breach of an express or implied warranty, negligence, and misrepresentation. In general, there is strict liability for such defective products, and the liability rests on those who are 'in the business of selling or otherwise distributing products' that sell or distribute a defective product whose defect harms individuals or property.[65] Berkowitz considers that the advent of prosumer 3D printing may result in more negligence cases regarding products because small-scale manufacturers may be more 'careless' in avoiding design defects.[66]

In the EU, liability for defective products is governed by two Directives from 1985 and 1999.[67] The 1985 Product Liability Directive (which has also served as the model for Australia's product liability law) appears to impose strict liability on commercial producers of defective products (and

---

[63] N. Freeman Engstrom (2013) '3-D Printing and Product Liability: Identifying the Obstacles' *University of Pennsylvania Law Review Online*, 162 (35) 35–41; N. D. Berkowitz (2015) 'Strict Liability for Individuals? The Impact of 3-D Printing on Products Liability Law' *Washington University Law Review*, 92(4), 1019–1053; H. Nielson (2015) 'Manufacturing Consumer Protection for 3-D Printed Products' *Arizona Law Review* 57(2), 609–622. Product liability is also briefly discussed in: L. Osborn (2014b) 'Regulating Three-Dimensional Printing: The Converging Worlds of Bits and Atoms' *San Diego Law Review*, 51, 553–621.

[64] Restatement (Third) of Torts: Product Liability § 2 (1998).

[65] Restatement (Third) of Torts: Product Liability § 1 (1998).

[66] Berkowitz (2015) 'Strict Liability for Individuals?', 1037.

[67] Namely: Council Directive 85/374/EEC of 25 July 1985 on the approximation of the laws, regulations, and administrative provisions of the Member States concerning liability for defective products [1985] OJ L210/29; and Council Directive 1999/34/EC of 10 May 1999 amending Council Directive 85/374/EEC on the approximation of the laws, regulations, and administrative provisions of the Member States concerning liability for defective products [1999] OJ L141/20.

suppliers of such products if the producer is not identified), and makes no distinction between types of suppliers or product defects.[68] However, producers are not liable if 'the product was neither manufactured by him for sale or any form of distribution for economic purpose nor manufactured or distributed by him in the course of his business'.[69] Indeed, the preamble to the Directive states that 'liability without fault should apply only to moveables which have been industrially produced'. Although terms such as 'for economic purpose', 'in the course of business', and 'industrially produced' are not precisely defined in the Directive, they might be thought to operate in a similar way to being 'in the business of selling or otherwise distributing products' for the purposes of US product liability as discussed above.

The Product Liability Directive has been transposed into Member States' domestic laws, such as the UK's Consumer Protection Act 1987. The defence in Article 7(c) of the Directive has taken the following form in section 4(c) of the Consumer Protection Act, that a defence to strict liability for a defective product is available in circumstances:

1. that the only supply of the product to another by the person proceeded against was *otherwise than in the course of a business of that person's*; and
2. that section 2(2) above does not apply to that person or applies to him by virtue only of things done *otherwise than with a view to profit* [emphasis added].

However, again no definition is given of what 'the course of a business' means precisely, nor of 'otherwise than with a view to profit'.

While worded slightly differently in the different jurisdictions, it seems in both the USA and EU that liability for defective products rests on whether the defendant is 'in the business of selling or otherwise distributing products' (USA), or distributing products 'for sale' or 'for economic purpose', 'in the course of business' or the products are 'industrially produced' (EU). This wording seems to involve products being produced on a certain scale ('industrial') and/or being supplied in the context of a sale or business. While this may not have been problematic previously, the small-scale production envisaged by prosumer 3D printing blurs the dis-

---

[68] J. Stapleton (2000) 'Restatement (Third) of Torts: Products Liability, an Anglo-Australian Perspective' *Washburn Law Journal*, 39, 363–403, 367–368.
[69] Directive 85/374/EEC, Article 7 (c).

tinction between industrial large-scale for-profit production (which would likely be covered by product liability laws) and individual small-scale private gifting (which would likely be beyond the scope of product liability laws).

Freeman Engstrom has considered that under US law, 'a plaintiff will have trouble prevailing in a PL action' against possible defendants since strict product liability applies only to commercial sellers and not occasional or casual vendors, and many hobbyist 3D printing prosumers' activity may well not fall within the definition of commercial seller.[70] Whether a particular individual and their activity fall within the definition of commercial seller will depend on the circumstances at hand. The manufacturer of the 3D printer which has been used to print a defective product would also not be liable, unless there was some defect with that printer itself which had been there since the printer left the manufacturer's possession and control. Furthermore, the 'digital designer', that is, 'the programmer who wrote the code that was fed into the printer to create the product at issue' also will not be liable since US product liability applies only to products, that is, tangible personal property—it seems that code will not qualify as such, so the digital designer is unlikely to be liable for a defective product emanating from the code they wrote.[71] Furthermore, if the 3D printing file has been distributed for no charge, for example, via Thingiverse, then the designer would most likely not be operating in the course of a business and so outside of the scope of US product liability for that reason as well.

Other potential defendants may be intermediaries such as the 3D printing design repositories or centralised sellers of 3D printed designs designed by third parties such as Shapeways, although they may attempt to escape liability in the USA for defects caused by defective products by defining themselves as 'service providers' rather than manufacturers.[72] Indeed, Shapeways frames itself as a service provider and excludes liability for defects in its terms and conditions.[73]

---

[70] Freeman Engstrom (2013) '3-D Printing and Product Liability, 36–37.

[71] § 19. Osborn argues that 3D printing design files may qualify as products in certain circumstances such as when they are mass-marketed, as opposed to custom-made files which may be considered to have a greater 'service' quality. See Osborn (2014) 'Regulating Three-Dimensional Printing', 568.

[72] Nielson (2015) 'Manufacturing Consumer Protection for 3-D Printed Products', 616.

[73] See: 'Limitation of Liability' in; Shapeways (2015) 'Terms and Conditions', http://www.shapeways.com/terms_and_conditions, accessed 12 September 2015.

This US-based discussion of 3D printing and product liability high-lighting the difficulties in applying existing tort law on product liability is likely to reverberate in other jurisdictions, such as the EU, in which definitions of manufacturing, distributing, and business/for-profit/commercial activities which may previously have been reasonably uncontroversial are challenged by small-scale prosumer 3D printing activity. In addition, even if the activity at hand does fall within the legal definitions for product liability, the difficulty may arise in enforcing such laws against small-scale producers which may be resident in another jurisdiction and may not be insured. The presumptions on which liability for defective products rests—that the manufacturer/distributor is a large enterprise able to shoulder the burden of liability—are also challenged by prosumer 3D printing activities.

In order to remedy some of these problems that prosumer 3D printing presents for US product liability, Berkowitz has suggested that a new category be created, of 'micro-sellers' which would cover those 'who surpass "occasional seller," but are not quite enterprise sellers', and instead of strict liability being applied to them, they would benefit from 'an equitable affirmative defense'.[74] This would involve a 'fairness' analysis—the defendant micro-seller would have to establish that strict liability, in fairness, should not apply, and the court could consider factors such as the defendant's experience in manufacturing, selling, and designing; the scale of their business; their ability to spread costs or buy insurance; the social desirability of the specific product at issue; and the defendant's good faith. Berkowitz identifies advantages to this defence over the current situation, including that only those defendants who are able to bear the cost and monitor for defects will be held liable, and that more sellers will be encouraged to take out insurance.[75] However, the defence would still be highly dependent on the specifics of the situation at hand, and so there may still be many marginal cases where it is unclear if a particular entity will qualify as a micro-seller, and if so, whether it should be held liable given the circumstances. Yet Berkowitz believes this defence strikes a balance between competing policy goals of promoting innovation while encouraging consumer safety.

However, Berkowitz's suggestion does not address the recurring problem of the enforceability of the law, in this case product liability laws. This is a problem that Lemley has noted for tort law in general, when faced

---

[74] Berkowitz (2015) 'Strict Liability for Individuals?', 1049.
[75] Berkowitz (2015) 'Strict Liability for Individuals?', 1051.

with production which is non-commercial and decentralised. Instead, he suggests that '[w]e may need to replace tort law with a social safety net as it becomes harder and harder to find those who make unsafe products and hold them liable'.[76]

This policy option of an increased social safety net may be more appropriate to the American context, where there is still no universal publicly funded healthcare, as it would give the victims of defective products free medical assistance with their injuries. Another option which may be more appropriate for countries such as those in Western Europe and Australia would be the route that New Zealand has taken to expand the social safety net with its Accident Compensation Scheme, in the process abolishing tort law for personal injury compensation. The Scheme provides financial compensation and support to those who have suffered personal injuries in New Zealand, and is funded through a combination of levies and government contributions.[77] Similar schemes might be considered in other jurisdictions rather than product liability laws, particularly if, as Lemley suggests, enforcing product liability laws will be rendered increasingly difficult by developments such as prosumer 3D printing.

Yet, these schemes deal with the consequences of defective products, providing an *ex post* remedy for those who have been injured, and do not address the cause of the problem, namely the defect causing the injury. In practice, thus, they may not provide the same deterrent effect on designers, manufacturers, and suppliers to ensure products are of a sufficiently high quality before they are passed on to consumers.[78] Furthermore, at a more philosophical or moral level, the New Zealand scheme does not entail that the tortfeasor 'takes responsibility' for her actions.[79] Accordingly, such schemes cannot be viewed as a panacea for the challenges that 3D printing presents to the product liability regimes.

[76] Lemley (2014) 'IP in a World Without Scarcity', p. 57.

[77] See: Oliphant, K. (2008) 'Accident Compensation in New Zealand: An Overview', in G. Schamps (ed.), *Evolution des droits du patient, indemnisation sans faute des dommages lies aux soins de sante: le droit medical en movement* (Brussels: Editions Bruylant).

[78] B. Howell (2004) 'Medical Misadventure and Accident Compensation in New Zealand: An Incentives-Based Analysis' *Victoria University of Wellington Law Review*, 35, 857–878.

[79] Enoch, D. (2014) '*Tort Liability and Taking Responsibility*' in J. Oberdiek (ed.), *Philosophical Foundations of the Law of Torts* (Oxford: Oxford University Press).

Medical Regulation

As mentioned at the beginning of this chapter, 3D printing has found many applications in the medical field. At one end of the spectrum lies the highly innovative and experimental printing of biological materials, usually using specialised 3D printers.[80] While inexpensive 3D printers have been used to print cells,[81] most bioprinting at the time of writing requires more sophisticated equipment which is outside the financial reach of the average prosumer—as well as expert knowledge in the field, which the average prosumer is also unlikely to have. The same is true of 3D printing pharmaceuticals—this is another medical application of 3D printing which at the time of writing is likely to be too complicated and/or too unaffordable for the average prosumer, even though desktop machines have been used to print medicines (albeit by skilled teams of researchers).[82] While the legal implications of bioprinting are beginning to be debated,[83] this discussion is currently beyond the prosumer focus of this book.

At the other end of the spectrum of medical uses for 3D printing lies the printing of prosthetics and other medical devices, which has already been embraced by a subsection of 3D printing prosumers (as well as medical research professionals). 3D printed prosthetics have obvious appeal inasmuch as they can be customised to the precise measurements and needs of each individual, and so far have presented cost savings from previous manufacturing methods. However, prosthetics are considered to be 'medical devices' in many jurisdictions, and specific regulation applies to them, to ensure among other aims that these devices meet minimum safety standards. This subsection will examine the relationship between this kind of prosumer printing of prosthetics and existing medical regulation.

[80] See: S. V. Murphy and A. Atala (2014) '3D bioprinting of tissues and organs' *Nature Biotechnology*, 32, 773–785.

[81] J. Leber (2013) 'A DIY Bioprinter Is Born', MIT Technology Review, http://www.technologyreview.com/view/511436/a-diy-bioprinter-is-born/, accessed 12 September 2015.

[82] See: S. A. Khaled, J. C. Burley, M. R. Alexander and C. J. Roberts (2014) 'Desktop 3D printing of controlled release pharmaceutical bilayer tablets' *International Journal of Pharmaceutics*, 461, 105–111; A. Goyanes, P. Robles Martinez, A. Buanz, A. W. Basit and S. Gaisford (2015) 'Effect of geometry on drug release from 3D printed tablets' *International Journal of Pharmaceutics*, in press.

[83] J. L. Tran (2015) 'To Bioprint or Not to Bioprint' *North Carolina Journal of Law and Technology*, 17, forthcoming; M. H. Park (2015) 'For a New Heart, Just Click Print: The Effect on Medical and Products Liability From 3-D Printed Organs' *Journal of Law, Technology and Policy*, 1, 187–210.

Some prominent 3D printing prosthetics projects have involved teams from the Global North travelling to developing countries in the Global South to 3D print items for people there who otherwise would not be able to access such prosthetics due to cost and availability barriers.[84] One of the most prominent projects in this area is e-NABLE, which describes itself as 'A Global Network of Passionate Volunteers Using 3D Printing To Give The World A "Helping Hand"'.[85] Participants in the web-based community network use their '3D printers, design skills and personal time to create free 3D printed prosthetic hands for those in need'.[86] At the time of writing, the e-NABLE website contains a suite of files which can be used to 3D print various upper-limb prosthetics; the files are all licensed using either Creative Commons or free software licences, with a clear disclaimer attached to the bottom of the page:

> By accepting any design, plan, component or assembly related to the so called 'e-NABLE Hand', I understand and agree that any such information or material furnished by any individual associated with the design team is furnished as is without representation or warranties of any kind, express or implied, and is intended to be a gift for the sole purpose of evaluating various design iterations, ideas and modifications. I understand that such improvements are intended to benefit individuals having specific disabilities and are not intended, and shall not be used, for commercial use. I further understand and agree that any individual associated with e-NABLE organization shall not be liable for any injuries or damages resulting from the use of any of the materials related to the e-NABLE hand.[87]

---

[84] See: B. Ouyang (2014) '3D Printing Low-Cost Prosthetics Parts in Uganda', Med Gadget, http://www.medgadget.com/2014/03/3d-printing-low-cost-prosthetics-parts-in-uganda.html, accessed 12 September 2015; D. Sher (2014) 'Kenya Based 3D Life Print Project Is Offering Mobile 3D Printing of Custom Prosthetics', 3D Printing Industry, http://3dprintingindustry.com/2014/12/08/3d-life-print-3d-printing-prosthetics/, accessed 12 September 2015; A. Leach (2014) '3D printed prosthetics: long-term hope for amputees in Sudan', The Guardian, http://www.theguardian.com/global-development-professionals-network/2014/jun/13/3d-printing-south-sudan-limbs, accessed 12 September 2015.

[85] Enabling the Future, http://enablingthefuture.org/, accessed 12 September 2015.

[86] Enabling the Future, 'Media FAQ', http://enablingthefuture.org/faqs/media-faq/, accessed 12 September 2015.

[87] Enabling the Future, 'Upper Limb Prosthetics', http://enablingthefuture.org/upper-limb-prosthetics/, accessed 12 September 2015.

e-NABLE envisages volunteers using their 3D printers to print out the hand prostheses contained within the design files on the site, with the volunteer printer bearing the cost of raw materials (which is estimated not to exceed US$200), and the website acting as a platform to unite volunteer printers with those in need of a prosthesis. At the time of writing, academic researchers at Creighton University are performing tests on e-NABLE's prostheses, although the website recommends 'careful observation while using these devices that involve your family physician's input and guidance' and that the prostheses not be used 'without consulting a physician prior to use and consult[ing] with them as to the best fit and use for you or the person you have created it for'.[88]

There is another page on e-NABLE's website specifically dedicated to safety guidelines for printing and using the prostheses, which includes various claims suggesting that the prostheses are of a lesser quality than prosthetics manufactured and supplied in a more traditional fashion:

- the prosthetics are not safe for the operation of heavy machinery, tools, equipment, and vehicles due to low grip strength development;
- they are not recommended for use by small children under three years of age as these children fall over a lot, may put the prosthesis or parts thereof in their mouths and may try to bite the device;
- they should not be exposed to high temperatures, which includes leaving them in a car on a hot day;
- it is implied they are not very durable and could 'break at any moment'.[89]

A further legal disclaimer is included along with the safety guidelines which set out the following:

- e-NABLE is a not-for-profit foundation which does not create the designs, does not manufacture or print the prosthetics, and does not certify that they operate properly or satisfy any regulatory requirements;
- the designs and prosthetics are conceptualised as 'gifts' from the foundation;

[88] Enabling the Future, 'FAQs (General)', http://enablingthefuture.org/faqs-general/, accessed 12 September 2015.

[89] Enabling the Future, 'Safety Guidelines', http://enablingthefuture.org/build-a-hand/safety-guidelines/, accessed 12 September 2015.

- the foundation does not make any representations or warranties regarding the designs and prosthetics and those designing the designs and printing the components are not contractors or employees of the foundation;
- the foundation and those associated with it (including volunteers) are released from any liability from any and all liability for acts or omissions— including negligent acts or omissions—causing damage, loss, injury, or death to the individual recipient from the use of the Design or the Component.[90]

Medical Device Regulation

As mentioned above, prosthetics are considered to be medical devices and so are subject to medical device regulation in many countries. In the USA, this entails regulation by the Food and Drug Administration's (FDA) Center for Devices and Radiological Health: all prosthetic devices must comply with federal regulations before they can be marketed within the USA.[91] There are three categories for devices: low risk (Class I), moderate risk (Class II), and high risk (Class III). External limb components are usually considered to fall within Class I as presenting low risk to their user and do not require pre-market notification to the FDA, but the devices must adhere to the FDA's 'general controls' to ensure their safety and effectiveness. These general controls requirements include *inter alia*: adhering to labelling standards; registration and device listing; Good Manufacturing Practices; and the keeping of records and reporting when the device causes or contributes to death or serious injury.[92] If a manufacturer does not adhere to the Current Good Manufacturing Practices, then a claim may lie in tort against that manufacturer, according to the decisions of various Circuit Courts.[93] However, a series of cases has entailed that most Class III devices (which require pre-market approval by the FDA—sometimes a lengthy process—before being introduced to the market) if defective do not usually give rise to a common law claim in negligence or product liability, which has resulted in courts subsequently 'consistently dismiss[ing]

---

[90] Ibid.

[91] See: L. Resnik, S. Klinger, V. Krauthamer and K. Barnabe (2010) 'U.S. Food and Drug Administration Regulation of Prosthetic Research, Development, and Testing' *Journal of Prosthetics and Orthotics*, 22(2), 121–126.

[92] See: U.S. Food and Drug Administration, 'General Controls for Medical Devices' http://www.fda.gov/MedicalDevices/DeviceRegulationandGuidance/Overview/GeneralandSpecialControls/ucm055910.htm#QSR, accessed 12 September 2015.

[93] Park (2015) 'For a New Heart, Just Click Print', 203.

patient claims against medical device manufacturers'.[94] Plaintiffs in such cases, to be successful, must plead that the manufacturer violated federal medical devices regulation and that these violations caused their injuries, which places a high evidentiary burden on plaintiffs.[95]

In the EU, medical devices were regulated at the national level until the 1990s, when the EU implemented its 'New Approach' to regulate market access and various other aspects of medical devices.[96] Under this framework, medical devices must comply with a list of 'essential requirements' and health and safety standards.[97] Those responsible for placing medical devices on the market must register with the competent authorities of the EU Member State in which the person has her registered place of business and provide a description of the devices concerned, and in the case that that person does not have a registered place of business in an EU Member State, then she must appoint an authorised representative in the EU.[98]

Medical devices regulation in the USA and EU envisages the existence of a centralised manufacturer or supplier of the device and that the device will be 'marketed'. The 3D printing of prosthetics by the e-NABLE community challenges all of these assumptions: the manufacture of the prosthetics takes place on a decentralised basis; it is not clear that the e-NABLE Foundation constitutes a manufacturer of these items since it plays the role of matching individuals seeking a prosthetic with individuals able to make one; and given the prosthetics are to be given as 'gifts' then it is unclear whether their publicising via the website actually constitutes

[94] D. Frank-Jackson (2011) 'The Medical Device Federal Preemption Trilogy: Salvaging Due Process for Injured Plaintiffs' *Southern Illinois University Law Journal*, 35, 453–497, 470.

[95] Frank-Jackson (2011) 'The Medical Device Federal Preemption Trilogy', 480.

[96] B. Lobmayr (2010) 'An Assessment of the EU Approach to Medical Device Regulation Against the Backdrop of the US System' *European Journal of Risk Regulation*, 1(2), 137–149. The New Approach for marketing of products was adopted in Council on 23 June 2008 and finally published in the Official Journal on 13 August 2008. It comprises two regulations and a decision by the European Parliament and Council: Council Regulation (EC) 764/2008 of 9 July 2008 laying down procedures relating to the application of certain national technical rules to products lawfully marketed in another Member State and repealing Decision No. 3052/95/EC [2008] OJ L218/21; Council Regulation (EC) 765/2008 of 9 July 2008 setting out the requirements for accreditation and market surveillance relating to the marketing of products, and repealing Regulation (EEC) No 339/93 [2008] OJ L218/30; Council Decision (EC) 768/2008/EC of 9 July 2008 on a common framework for the marketing of products, and repealing Council Decision 93/465/EEC [2008] OJ L218/82.

[97] Council Directive 93/42/EEC concerning medical devices [1993] OJ L169/1 ('Medical Devices Directive'), Article 3.

[98] Medical Devices Directive, Article 14.

'marketing' of the device. However, in European medical device regulation, 'manufacturer' and 'placing on the market' are specifically defined terms.[99] 'Manufacturer' is the person or organisation 'with responsibility for the design, manufacture, packaging and labelling of a device before it is placed on the market under his own name' but the obligations incumbent on a manufacturer are not applicable to someone who 'assembles or adapts devices already on the market to their intended purpose for an individual patient'. 'Placing on the market' is defined as 'the first making available in return for payment or free of charge of a device... with a view to distribution and/or use on the Community market'. This would seem to entail that e-NABLE itself may be conceptualised as a manufacturer for the purposes of EU medical device regulation, and even if it prohibits charging for the prosthetics, it is still placing them on the market. However, an individual volunteer printing out an e-NABLE prosthetic for a specific recipient is likely to be someone 'who assembles or adapts devices already on the market... for an individual patient' and so not subject to the obligations incumbent on a manufacturer.

Medical devices regulation in the EU is currently undergoing a process of revision, with a new regulation which may be introduced in 2016. However, the revision has already been criticised by Vollebregt for being insufficiently equipped to regulate custom devices that are manufactured by 3D printing since customised devices will still be subjected to a low regulatory burden (even if complex 3D printed customised devices may pose high risks).[100] A model to follow might be found in the proposals relating to the regulation of in vitro diagnostics (which test samples of tissue or bodily fluid) which are manufactured and used only within a single health institution,[101] placing quality safety requirements on such device manufacture and ensuring that there is a controlled production environment.[102] However, this provision still envisages in vitro diagnostic medical devices being manufactured within 'a single health institution', which may not be the case at all for the manufacture of, for instance, e-NABLE prosthetics.

[99] Medical Devices Directive, Article 1(f) and Article (1) (h).

[100] E. Vollebregt (2014) '3D printing of custom medical devices under future EU law', Medical Devices Legal, http://medicaldeviceslegal.com/2014/03/05/3d-printing-of-custom-medical-devices-under-future-eu-law/, accessed 12 September 2015.

[101] Proposal for a Regulation of the European Parliament and of the Council on in vitro diagnostic medical devices, and amending Directive 2001/83/EC, Regulation (EC) No. 178/2002 and Regulation (EC) No. 1223/2009, COM(2012) 542 final, Article 4(5).

[102] Vollebregt (2014) '3D printing of custom medical devices under future EU law'.

Product Liability

Even if e-NABLE or one of its volunteers could be conceptualised as supplying prosthetics which were not in compliance with medical device regulation, and if one such prosthetic caused harm by being defective, then tortious product liability may be engaged. However, the discussion above on product liability in both the EU and USA suggests that if the product is not 'sold' in the course of a 'business' then these laws would seem not to apply.

In any event, in the USA at least, a claim would also have to show that 'the foreseeable risks of harm posed by the drug or medical device are sufficiently great in relation to its foreseeable therapeutic benefits that reasonable health-care providers, knowing of such foreseeable risks and therapeutic benefits, would not prescribe the drug or medical device for any class of patients'.[103] In practice, this usually involves a case-by-case risk–benefit analysis but some US courts have precluded liability by concluding that all medical products should be viewed as unavoidably unsafe.[104] It is thus possible that an American court, even if it does accept that liability may exist for a defective prosthesis, may find that the therapeutic benefits of e-NABLE prostheses outweigh the risks of harm that the prosthetics pose.

Negligence

Negligence liability may also be engaged by a prosthetic which causes the user harm. Essentially, for such liability to be established, three elements must be found: a duty of care; a breach of that duty; and damage caused by that breach. In this scenario, possible actors which owe that duty of care could be e-NABLE itself; the volunteer who used their 3D printer to create the prosthetic; and a medical professional who may be involved with prescribing/overseeing the use of the prosthetic.

The current test for establishing a duty of care in English law (with similar tests employed in other common law jurisdictions) is: whether the defendant should have foreseen harm to the claimant; secondly, whether there was a relationship of proximity between the claimant and the defendant; and, thirdly, whether it is fair, just, and reasonable that the defendant should owe a duty of care to the claimant.[105]

---

[103] Restatement (Third) of the Law of Torts: Products Liability § 6(c).

[104] Y. J. Lu (2010) The Change in Knowledge Proposal: Repairing Preemption Doctrine in Medical Products Liability, SSRN Working Paper, http://ssrn.com/abstract=1957954, accessed 12 September 2015, 37.

[105] *Caparo Industries plc v Dickman* [1990] 2 AC 605, 617–618, HL, *per* Lord Bridge.

It would seem that e-NABLE and the volunteer ought to have foreseen that their conduct in designing and making available design files for 3D printed prosthetics and in printing one of these prosthetics, respectively, could cause harm to the recipient of the prosthetic. The relationship between these parties is also likely to be considered sufficiently 'proximate' for the purposes of the duty of care test. The third prong of the test, whether it is fair, just, and reasonable that the defendant should owe a duty of care to the claimant, may be more debatable, although in practice there are limited cases in which it has not been considered fair, just, and reasonable to impose such a duty.[106] It is possible in these scenarios that public policy considerations may be raised to negate a duty of care, such as the value in having these prosthetics more accessible to the community than previously.

If a duty of care is established, then in order to show that it has been breached due to negligence, there must be a failure to take the care expected by the community for the activity in question. If the defendant professes to have particular skills, then she is judged by the standard which would have been exercised by a reasonable person with this knowledge and skills, even if she does not actually have this expertise. What taking reasonable care means in particular circumstances is subject to four considerations: the likelihood that the activity in question will cause damage; the likely severity of the damage if it occurs; the difficulty and expense of averting the danger; and, when appropriate, the value to society of the activity undertaken.[107] These considerations can be used to judge the factual scenario to determine whether reasonable care was taken. Out of the four, the value to society of the activity undertaken, that is the low-cost 3D printing of customised prosthetics, may be the consideration which weighs most, if the value of the activity does not negate a duty of care arising in the first place as discussed above.

If a duty of care is established, and it is also established that reasonable care was not taken in the printing of the prosthetic and/or the making available of the 3D printing design file, then it would be for the claimant to prove that the defendant's resulting breach of duty caused the damage. It is not necessary to show that the defendant was the sole or even major cause of the damage, but the claimant must prove that 'but for' the

---

[106] Nolan, D. and Davies, J. (2013) 'Torts and Equitable Wrongs' in A. Burrows (ed.), *English Private Law* (Oxford: Oxford University Press), p. 939.

[107] Nolan, D. and Davies, J. (2013) 'Torts and Equitable Wrongs', p. 944.

defendant's negligence, the damage would not have happened. It is also possible for there to be multiple defendants if each can be shown to have owed a duty of care which was breached and which caused the negligence.

As mentioned above, at various points in the literature on e-NABLE's website the intervention of medical professionals is mentioned. If a medical professional recommends and oversees the use of a defective 3D printed prosthetic, then a negligence medical malpractice case may lie against her, in which the elements of negligence must be established.[108] For medical professionals, the proper standard of care in most jurisdictions will be beyond that expected from the ordinary lay reasonable person, based on the additional skills and expertise the medical professional can be presumed to have. As with other negligence scenarios, whether the proper standard of care has been complied with will depend on the circumstances of the case at hand. In the State of Illinois for instance, the 'learned intermediary doctrine' does require medical doctors who prescribe a medical device to warn their patients of the dangers of the medical device.[109] Thus, if a medical professional in Illinois oversees the use of an e-NABLE prosthetic without warning their patient of the dangers and risks posed by the device, then she may be considered not to have exhibited the proper standard of care.

Any defendants may wish to avail themselves of the consent defence to negligence liability, *volenti non fit injuria* ('no wrong is done to one who consents'), by arguing that the recipient of the prosthetic has consented to an act which otherwise would be a tort and/or has agreed to assume the risk of injury as per the disclaimers on the e-NABLE website detailed above, which purport to release e-NABLE and its volunteer printers from any liability, including for negligence, for acts or omissions causing damage, or injury to recipients of the prostheses. However, in English law at least, there are various statutory provisions which invalidate consent, notably: section 7 of the Consumer Protection Act 1987 which prevents the limitation or exclusion of liability for defective products imposed by the Act by any contract term, notice, or other provision; and section 2 of the Unfair Contract Terms Act 1977 which provides that a person in the course of a business cannot, by reference to any contract term or notice, exclude or restrict her liability for death or personal injury resulting from negligence.

---

[108] Park (2015) 'For a New Heart, Just Click Print', 206.
[109] *Hansen v. Baxter Healthcare Corp.*, 764 N.E.2d 35, 42 (Ill. 2002).

Although, of course, it is at best debatable whether what e-NABLE and its volunteers are doing could be classified as a business activity, if the prosthesis is a defective product, then it would seem that at least under English law e-NABLE and the volunteer printer could not rely on the disclaimers made on e-NABLE's website to exclude their liability for negligence.

Conclusion

This chapter has explored the challenges 3D printing brings by enabling the creation of objects that are dangerous or otherwise undesirable. The peculiar case of the Liberator certainly poses legal headaches in the USA, given the right to bear arms enshrined in the American Constitution, along with the expansive right to free speech which has been subject to a very broad interpretation. The ongoing constitutional challenge to the US authorities' attempts to restrict the Liberator will provide some interesting answers to the extent to which these constitutional rights protect the distribution of information about how to make one's own gun.

Slightly more mundane problems posed by 3D printing are the creation of objects which do not conform to product safety standards, and the questions of liability. This is illustrated in particular with the e-NABLE project which aims to increase access to custom-made prosthetics but also provokes difficult legal questions around medical regulation and product safety.

Yet much of this discussion is very theoretical, both for 3D printed weapons and for more banal products, since at the time of writing there does not appear to be widespread manufacture of these items by 3D printers. The remarks from the University of Toronto team which attempted to make a non-functioning Liberator demonstrate the difficulty in using a 3D printer, especially one oriented to the consumer market, to print such items. In addition, there do not appear to be actual instances of defective 3D printed products causing damage or injury. While, again, there is very much the potential for 3D printing these objects to become widespread in the future, at the present time this is not the case. In principle, 3D printing does challenge the areas of law mentioned in this chapter, but in practice it seems the challenge at the moment is minimal.

CHAPTER 4

Selfies in Another Dimension:
The Implications of 3D Scanning

Abstract The rise of 3D printing has been supplemented by the increasing availability of 3D scanning techniques, whereby data is collected from pre-existing objects or even people and turned into a virtual 3D model which can then be used as a design file for input into a 3D printer, then to be printed as a 3D object. This innovation of 3D scanning greatly expands the potential content for 3D printers, but also provides challenges to various areas of law, namely IP and data privacy, since the 3D scanning of human bodies is an initial consumer-oriented application of this technique.

Accordingly, this chapter will continue the debate from Chap. 2 on IP and piracy, given the likely proliferation of content for 3D printers as a result of 3D scanning, as well as new possibilities to 'reverse engineer' inventive objects which trigger fresh IP concerns. However, the chapter will go beyond the discussion in Chap. 2 by introducing issues of privacy and surveillance in 3D printing. Legal ownership of this scanned data will be explored, as well as the applicability of any personality or publicity rights to individuals' scanned images. The core area of law which will be considered here, though, will be privacy and data protection, and the extent to which these rights and interests might apply to this practice of digitising people's bodies via 3D scanning. Previous experience with the legality of other types of biometric data will be assessed, along with the guidance it may provide to how 3D scanning and the data it produces should be managed.

© The Editor(s) (if applicable) and The Author(s) 2016   81
A. Daly, *Socio-Legal Aspects of the 3D Printing Revolution*,
DOI 10.1057/978-1-137-51556-8_4

INTELLECTUAL PROPERTY AND REVERSE ENGINEERING

3D scanners bring another 'dimension' to the interaction of 3D printing and IP laws and enforcement since they open up the possibility of scanning pre-existing objects, from which a CAD file can be created, containing a blueprint for the object which can then be printed using a 3D printer. This process, by which an object is 'taken apart' physically or conceptually to see how it works in order to recreate that object, is known as 'reverse engineering' and raises fresh concerns about IP infringement, as well as amplifying those concerns already discussed in Chap. 2.

The legality of reverse engineering has been a topic of discussion before the advent of 3D printing and 3D scanning. Samuelson and Scotchmer have considered that reverse engineering in traditional manufacturing industries has generally been a lawful way to acquire knowledge about manufactured products, since the innovator of such a product has been protected by the cost of, and time taken to, reverse engineer a product.[1] Furthermore, in US law, publishing information which has been learned through lawful reverse engineering has also been deemed legal, although the DMCA has restricted such publication vis-à-vis TPMs, and other jurisdictions such as the EU have sought to restrict publication in certain circumstances, such as the Software Directive's prohibition on publishing information gained through decompiling software programs to achieve interoperability.[2] Of course, 3D scanning greatly reduces the cost and time necessary to reverse engineer an object, and so may give rise to calls to restrict the practice in order to protect the interests of the designers and manufacturers of such objects.

This section will examine some of the IP aspects of 3D scanning, looking at both whether the scans themselves can be protected by IP, and whether the act of scanning may infringe the IP of others.

Copyright

In US law, Weinberg considers that a scan of a physical object is not 'independently' protected by copyright since it is not sufficiently 'original' to merit copyright protection, entailing that 'useful objects' for the purposes of US

---

[1] P. Samuelson and S. Scotchmer (2002) 'The Law and Economics of Reverse Engineering' *Yale Law Journal*, 111(70), 1575–1664.

[2] Samuelson and Scotchmer (2002) 'The Law and Economics of Reverse Engineering'.

copyright law will not attract copyright protection, nor will their 3D scans.[3] While scanning objects may be considered analogous to taking a photograph (which may attract copyright protection if sufficiently creative), it is unlikely that elements of creativity—such as posing the subject or adjusting the lighting—will be found in the act of scanning. Indeed, Osborn considers that 'utilitarian' 3D scans of objects are more akin to the kinds of photographs not held to attract copyright protection in US case law, although altering a CAD file created from a 3D scan may constitute a sufficient element of creativity to entail that the final version of the file is indeed protected by copyright.[4]

Scans of 'creative' objects will not in themselves attract copyright protection, but they are representations of physical objects which are protected by copyright, with the consequence that anyone wishing to scan such an object should seek permission from the holder of the copyright over that object since scanning the object makes a copy of that object.[5] Accordingly, the copyright holder of the original creative object—and not the person who has conducted the scan—must give permission for the file created by the 3D scan to be copied or otherwise distributed.

In UK copyright law, Bradshaw et al. have considered the status of CAD files created from a 3D scan of an object for which there originally was a design document created before the object was created. In this discussion, section 51 of the CDPA 1988 should be borne in mind from Chap. 2, which provides that the copyright in a 'design document' is not infringed by making an article from it, and that 3D printing design files will fall within the definition of design document. Case law suggests that any copyright in the original design document for the object is not infringed by the creation of a 3D printing design file using a 3D scanner, although other rights such as design rights may be infringed by the creation of a 3D scan of the object intended to be used with a 3D printer (although if the uses are private and non-commercial, then design rights under UK law are unlikely to be infringed).[6]

---

[3] Weinberg (2013) *What's The Deal With Copyright and 3D Printing*, pp. 15–16, citing *Meshwerks v Toyota Motor Sales*, 528 F.3d 1258 (10th Cir. 2008) and *Bridgeman Art Library v Corel Corporation*, 25 F.Supp 2d 421 (S.D.N.Y. 1987), modified 36 F.Supp.2d 191 (S.D.N.Y. 1999).

[4] L. Osborn (2014a) 'Of PhDs, Pirates, and the Public' *Texas A&M Law Review*, 1, 811–835, 831.

[5] Weinberg (2013) *What's The Deal With Copyright and 3D Printing*, p. 18.

[6] Bradshaw, Bowyer and Haufe (2010) 'The Intellectual Property Implications of Low-Cost 3D Printing', 25, discussing *BBC Worldwide and Anor v Pally Screen Printing and others* [1988] FSR 665 and *Mackie Designs v Behringer Specialised Studio Equipment and others* [1999] RPC 717.

But does the design file created by scanning an object attract copyright protection in UK law? Various authors have considered this question, particularly on the point of whether such a reproduction is sufficiently 'original' to attract copyright protection in itself. Li et al. note that the historical position in UK law has been that a work must not be copied but 'no more than skill, judgement or labour needed to be expended in its creation'; however, recent case law from the CJEU suggests that the originality requirement is comprised by the work being the author's 'intellectual creation'.[7] This appears to involve some element of creative freedom being expressed in an original manner, and will not be fulfilled by creation dictated by technical considerations, rules, or constraints which do not permit this freedom.

Thus, if the act of scanning an object is viewed as not involving some element of creative freedom on behalf of the person performing the scanning, then it is unlikely to attract copyright protection. Mendis has argued, based on *Antiquesportfolio* (where photographs of antiques were held to be original copyright works because of the positioning of the object and composition of the photograph)[8] and *Painer* (in which the CJEU discussed similar elements as expressing a photographer's free and creative choices),[9] that design files derived from a 3D scan of another object may attract a new copyright depending on the level of skill, effort, and judgement used when scanning that object, which might be fulfilled by, for example, creative choices such as selecting particular views of the object to be scanned.[10] Yet, consumer-oriented 3D scanners seem to require minimum or no creative input from the person placing the object within them—for instance, MakerBot's Digitizer scanners require minimal human effort of any kind to carry out the scan, aside from placing the object on the machine's rotating platform. It would be difficult to describe such actions as constituting a sufficiently creative choice to create a new copyright in the digital design created in the file which comes about as a result of the scanning. In any event, under current UK law, 3D digital models of works of artistic craftsmanship which are manufactured on an industrial scale will not attract copyright protection as a result of the *Lucasfilm* decision discussed in Chap. 2.[11]

---

[7] P. Li, S. Mellor, J. Griffin, C. Waelde, L. Hao and R. Everson (2014) 'Intellectual property and 3D printing: a case study on 3D chocolate printing' *Journal of Intellectual Property Law & Practice*, 9(4), 322–332.

[8] *Antiquesportfolio.com plc v Rodney Fitch & Co Ltd* [2001] FSR 23.

[9] Case C-145/10 *Painer v Standard Verlags GmbH* 30 [2012] ECDR 6 (ECJ (3rd Chamber)).

[10] Mendis (2014) '"Clone Wars II": Episode II', 278.

[11] Mendis (2014) '"Clone Wars II": Episode II', 278–279.

However, if the object being scanned is itself protected by copyright, then so long as a substantial part of it is being copied using, for example, 3D scanning, it will still be an infringement even if the size changes, dimensions are altered, elements of the original work are left out, or parts added.[12]

Patents

In UK law, scanning an item which is patented to make a 3D printing design file, and then printing out that item will infringe the patent, but there are defences to infringement if the object is made privately and for non-commercial purposes according to section 60(5) of Patents Act 1977.

If the file is uploaded onto a site such as Thingiverse, in UK patent law, it might fall foul of the provision in section 60(2) that prohibits supplying others with the means to infringe a patent, but not clear from case law that a 3DP design file constitutes such 'means',[13] and it is not relevant if the person uploading the file intends for others to use it for private and non-commercial purposes.

Design Rights

Scanning an object which is protected by a design right will constitute 'copying'. As discussed in Chap. 2, the exceptions for design right infringement are of great importance here. Given copying design rights for non-commercial and private purposes in UK law is an exception to infringement for registered design rights, and copying for non-commercial purposes is an exception to the infringement of unregistered design rights, many instances of scanning objects protected by design rights will not be infringing.

Trade Marks

As regards trade marks and 3D scanning, again the discussion in Chap. 2 is relevant here. Essentially, it would seem that scanning an object which either is a 3D shape mark or contains a trade marked item on its surface will not infringe the trade mark owner's rights if the scanning is not a 'use in commerce' for the purposes of trade mark law. For those jurisdictions

---

[12] Li, Mellor, Griffin, Waelde, Hao and Everson (2014) 'Intellectual property and 3D printing: a case study on 3D chocolate printing'.

[13] Bradshaw, Bowyer and Haufe (2010) 'The Intellectual Property Implications of Low-Cost 3D Printing', 27.

where trade mark dilution is recognised, scanning such an object and then printing it out on a 3D printer may contribute to the dilution of that mark.

Scanning in Practice

Thus, as can be seen from the preceding discussion, 3D scanning does introduce further complexity to the relationship between 3D printing and IP. However, it is not clear at the time of writing how widespread scanning is, especially among consumers or prosumers.

Indeed, Reeves and Mendis identify one main barrier to wider consumer adoption of 3D scanning—the fact that the design files produced by scanning are not of a high quality such that they cannot achieve an exact digital replica of the original object, and so this was unlikely to cause much harm to IP owners.[14]

Again, in principle, scanning does implicate IP infringements, and these infringements may become more of a challenge as scanning technology improves. However for the moment, IP owners do not seem particularly threatened by its existence.

Privacy and Surveillance

As mentioned in the introductory part of this chapter, the socio-legal implications of 3D scanning go beyond IP concerns, particularly since an initial use of 3D scanners has been in fashion to collect data on human bodies in order to customise clothing. The privacy and data protection concerns relating to this data are paramount.

One example of 3D scanning in consumer-oriented fashion applications is the Australian-based mPod, which claims to be the world's first fully automated 3D body scanner.[15] These mPods have been installed in various shopping malls in large Australian cities. A person enters an mPod, essentially a scanning booth, which captures their body measurements and then sends the mapping data to certain fashion partners to recommend the correct size of clothing for that individual. A further feature is that the body measurements are also used to calculate BMI, body composition, and hip-to-waist ratio, giving that an individual a 'myBody report' on the

---

[14] Reeves and Mendis (2015) 'The Current Status and Impact of 3D Printing Within the Industrial Sector', pp. 45–46.

[15] mPort, 'FAQ', http://mport.com.au/home/faq, accessed 12 September 2015.

state of their health. They can profile how their body is changing over time by comparing this data with data gained from previous visits to the mPod.

The company behind mPod, mPort, does include a Privacy Policy on its website, which acknowledges that, being based in Australia, it is governed by the Privacy Act 1988 (Cth) and the Australian Privacy Principles (APPs). mPort details the kinds of personal information it collects in the course of its scanning business:

- contact and identification information such as your name, address, telephone number, email address, date of birth, gender and age;
- physical measurements, including height, waist, bust, hip, weight over time as you scan; history of when and where you scan yourself;
- preferences or information you disclose on our website for our various applications;
- sensitive information including:
  - health information: where you consent to the collection of such information and if relevant to the provision of mPort's services; and
  - where you are applying for a position with mPort, criminal record information where relevant; and
- other information required for mPort's functions and activities.[16]

These types of information and data gathered by the mPort system will fall within the definition of 'personal information' in section 6 of the Privacy Act ('information or an opinion about an identified individual, or an individual who is reasonably identifiable'), and some of this information will qualify as 'sensitive information' since it is either 'health information', 'genetic information', 'biometric information', or 'biometric templates'. The use of body scanning technology to collect physical body measurements may also result in the collection of other information about that individual's body, as well as the analysis of those measurements giving rise to other information or opinions about that individual, such as their race or ethnicity and possibly also socio-economic class.[17]

---

[16] mPort, 'Privacy Policy', http://mport.com.au/home/privacypolicy, accessed 12 September 2015.
[17] See: J. C. K. Wells, T. J. Cole, D. Bruner and P. Treleaven (2008) 'Body shape in American and British adults: between-country and inter-ethnic comparisons' *International Journal of Obesity*, 32, 152–159.

Such personal information must be handled in accordance with the APPs, which contain *inter alia* the following obligations:

- Entities handling personal information must have a privacy policy concerning how it manages the personal information it handles;[18]
- Individuals must have the option of not identifying themselves, or of using a pseudonym, when dealing with an entity in relation to a particular matter (except when it is 'impracticable' for the entity to deal with individuals who have not identified themselves or who have used a pseudonym);[19]
- Entities must only collect personal information if this collection is 'reasonably necessary' for its functions;[20]
- Sensitive information must only be collected if the individual consents to this;[21]
- Personal information collected for a particular purpose in general should not be used or disclosed for another purpose;[22] and
- Transfers of personal information to overseas recipients must only occur by ensuring that the recipient is also bound by the APPs.[23]

As already mentioned, mPort acknowledges the application of Australian privacy law to its activities, conforming already to the requirement that it has a privacy policy. However, this example of mPort's 3D scanning technology and the kind of information it collects is illustrative of the deficiencies of Australian privacy law in particular, and data privacy laws more generally, for the following reasons.

Firstly, the 3D scanning of bodies has broader implications than the identification of the correct set of laws which apply to the data generated. This scanning and digitisation of bodily information contributes—alongside developments such as Quantified Self-tracking,[24] cloud

---

[18] Australian Privacy Principle 1.3.
[19] Australian Privacy Principle 2.1–2.2.
[20] Australian Privacy Principle 3.1–3.2.
[21] Australian Privacy Principle 3.3.
[22] Australian Privacy Principle 6.1.
[23] Australian Privacy Principle 8.1.
[24] A. Daly (2015) 'The Law and Ethics of "self-quantified" health information: an Australian perspective' *International Data Privacy Law*, 5(2), 144–155.

computing,[25] locative media,[26] and the Internet of Things[27]—to the proliferation of data and information being generated by and about individuals, which ultimately has serious implications for their privacy. This further method of creating, documenting, and tracking information about individuals contributes to the pervasive 'surveillance society'[28] in which citizens increasingly find themselves. Furthermore, the fact that 3D scanning creates a particularly intimate picture of an individual's body renders this information especially sensitive biometric data, whose creation and collection, Clarke argues, has 'extremely serious implications for human rights in general, and privacy in particular'.[29]

Related to this point is whether the data collected through services such as mPort's is passed on to other parties, and if so, the identity of those parties receiving the data. mPort's own Privacy Policy states that it 'holds, uses and discloses' users' personal information 'where it is reasonably necessary' for providing customers with its services, informing customers about mPort's products and services and those of 'selected third parties', assessing employment applications and 'any other legal requirements'.[30] In addition:

> mPort may also use your personal information for purposes related to the above purposes and for which you would reasonably expect mPort to do so in the circumstances, or where you have consented or the use is otherwise in accordance with law.

---

[25] P. De Filippi and S. McCarthy (2012) 'Cloud Computing, Centralization and Data Sovereignty' *European Journal of Law and Technology* 3(2); D. Svantesson and R. Clarke (2010) 'Privacy and Consumer Risks in Cloud Computing' *Computer Law & Security Review*, 26(4), 391–397.

[26] Meese, J. (2014) 'Google Glass and Australian Privacy Law: Regulating the Future of Locative Media' in R Wilken and G Goggin (eds.), *Locative Media* (London: Routledge); A. De Souza e Silva and J. Frith (2010) 'Locational Privacy and Public Spaces: Media Discourses on Location-Aware Mobile Technologies' *Communication, Culture & Critique*, 3(4), 503–525.

[27] S. R. Peppet (2014) 'Regulating the Internet of Things: First Steps Toward Managing Discrimination, Privacy, Security & Consent' *Texas Law Review*, 93, 85–176.

[28] G. T. Marx (1985) 'The Surveillance Society', *The Futurist*, 19(3) 21–6; D. H. Flaherty (1988) 'The Emergence of Surveillance Societies in the Western World: Toward the Year 2000' *Government Information Quarterly*, 5(4), 377.

[29] R. Clarke (2001) 'Biometrics and Privacy', http://www.rogerclarke.com/DV/Biometrics.html, accessed 12 September 2015.

[30] mPort, 'Privacy Policy'.

There is significant vagueness in this language, such that it would seem mPort can disclose personal information concerning its users in various circumstances. The recipients of such disclosures would seem to include not only other private companies but also law enforcement agencies (mPort may be forced to disclose personal information to comply with 'legal requirements' such as user data access requests). This kind of wording is fairly common in companies' privacy policies. What then happens to such data when it is passed onto such other parties, and the subsequent profiling of individuals using this data matched with other data sets, has given rise to grave concerns over the lack of transparency regarding these activities and a lack of agency on behalf of individuals.[31]

The law on data privacy in Australia is similar to the EU's Data Protection Directive,[32] although the EU has considered the Australian legislation to be 'inadequate'.[33] However, even the EU's more robust laws still do not provide complete protection of individual privacy in these circumstances. EU data protection law allows personal data to be collected in certain scenarios: where unambiguous consent has been given, where necessary for performing a contract, where necessary to comply with a legal obligation, where necessary to protect the vital interests of the individual, where necessary for performing a task in the public interest, or where necessary for the 'legitimate interests' pursued by the entity collecting the data.[34] This is similar to the bases on which personal information can be collected in Australian law, as mentioned above. In addition, the level of fines that can be imposed on the finding of a data protection breach are so low in amount and not always enforced that large firms may find it more profitable to breach the laws and pay the fines rather than follow the law in the first place.[35]

---

[31] F. Pasquale (2015) *The Black Box Society: The Secret Algorithms That Control Money and Information* (Cambridge: Harvard University Press).

[32] Council Directive 95/46/EC of 24 October 1995 on the protection of individuals with regard to the processing of personal data and on the free movement of such data [1995] OJ L 281/31 ('Data Protection Directive').

[33] Greenleaf, G. (2008) 'Privacy in Australia' in J. B. Rule and G. Greenleaf (eds.), *Global Privacy Protection: The First Generation* (Cheltenham: Edward Elgar), p. 167.

[34] Data Protection Directive, Article 7.

[35] P. Ducklin (2013) 'How effective are data breach penalties? Are ever-bigger fines enough?', Nakedsecurity, http://nakedsecurity.sophos.com/2013/04/26/how-effective-are-data-protection-regulations/, accessed 12 September 2015.

The creation and storage of data 'relating to the private life of an individual' has been recognised by the European Court of Human Rights as amounting to an interference with an individual's privacy rights under Article 8 of the European Convention on Human Rights (ECHR).[36] What was at issue in this case was the collection and holding of DNA samples from individuals who were arrested in the UK but later acquitted or against whom charges were dropped, and the European Court found that this was a violation of the individuals' right to privacy. While the creation and storage of such data relating to the private life of an individual can be seen as a *prima facie* interference with that individual's privacy, Article 8 of the ECHR provides that such an interference can be justified if it is 'in accordance with the law and is necessary in a democratic society'.

In the EU, social network Facebook's utilisation of facial recognition software vis-à-vis photos uploaded to the site without the explicit opt-in consent of users was deemed by Hamburg's data protection regulator to be incompatible with EU data protection law.[37] Facebook subsequently disabled facial recognition features for its users in the EU. The Article 29 Working Party, the independent European advisory body on data protection and privacy, has considered facial recognition software among other developments in biometrics. In general, it considers that biometrics' potential impact on individual privacy is high because of the ways in which they permit automated tracking, tracing, and profiling.[38] Specifically regarding facial recognition, the Working Party considers that entities wishing to use such software on images of individuals must specifically inform those individuals that facial recognition will be used, and these individuals must have an option as to whether they consent to this happening.[39] A user's acceptance of the service's overall terms and conditions will usually not be sufficient for consent to the use of facial recognition software.[40]

---

[36] *S and Marper v United Kingdom* [2008] ECHR 1581, 67.

[37] Information Age (2011) 'Facebook facial recognition breaks EU law – regulator', http://www.information-age.com/technology/security/1669438/facebook-facial-recognition-breaks-eu-lawDOUBLEHYPHEN-regulator, accessed 12 September 2015.

[38] Article 29 Working Party, *Opinion 03/2012 on developments in biometric technologies* (WP193).

[39] Article 29 Working Party, *Opinion 02/2012 on facial recognition in online and mobile services* (WP192), p. 6.

[40] Article 29 Working Party, *Opinion 02/2012*, p. 8.

Turning back to mPort's service, its terms and conditions as they stand would likely be insufficient to ensure compliance with EU data protection law if mPort was using facial recognition software on the scanned images of individuals' bodies which are captured when its 3D scanning booth is used. Bunn has considered that photographs of identifiable individuals will constitute biometric information which is likely to be considered 'sensitive information' for the purposes of Australian privacy law; she has also viewed that the use of facial recognition technology on such images and the creation of further data from about the images ('metadata') may create compliance risks regarding user consent for an organisation carrying out these activities under general terms and conditions.[41] If mPort is indeed using software such as facial recognition on the scanned body images, then in order to ensure it is fully compliant with Australian privacy law, it may wish to update its Privacy Policy and other Terms of Use to make this explicit and ensure it seeks informed user consent. However, in the absence of a regulatory scenario similar to what happened to Facebook in the EU, the position in Australia regarding specific consent to such activities remains unresolved.

A supplementary issue regarding body scans involves ownership of the image created. The situation with photographs is that, in the absence of any contractual agreements, the first owner of the copyright is normally the person who creates the artefact, namely the photographer. In Australian copyright law, in certain circumstances, this is not the case, such as where the photograph was taken by a freelance photographer for a client, in which case the client is considered the first owner of the copyright. While it could be argued that body scan images are analogous to photographs, it would likely be too difficult to argue that an entity such as mPort is analogous to a freelance photographer, in which case this ownership scenario would be unlikely to come to pass. In any event, mPort in its Terms and Conditions asserts ownership of IP over any IP created in its scanning pods which would encompass IP ownership over the body scan images.

Notwithstanding the general rule that the first owner of copyright is usually the artefact's creator, some jurisdictions have recognised rights of publicity, or personality rights, which allow an individual to control the commercial use of her name, image, and likeness. In practice, such

---

[41] A. Bunn (2013) 'Facebook and face recognition: kinda cool, kinda creepy' *Bond Law Review*, 25(1), 35–69.

publicity rights are mainly useful for celebrities, whose images, more than those of general members of the public, may actually have some commercial value to be exploited. The idea of these rights as being proprietary and having monetary value has been principally developed in US jurisprudence, while certain civil law jurisdictions such as France and Germany have for some time protected name and image as an element of personal privacy.[42] Publicity rights in the USA are protected by state-level laws and so the scope of the right and protection varies from state to state; however, the right would usually be infringed by unauthorised commercial use of a person's identity to attract attention to a product or advertisement.[43]

In the situation at hand, it may be difficult firstly to establish that publicity rights exist, at least where the body scan is of an individual who is not a celebrity or otherwise publicly well known, since there may be no commercial value to be exploited in the use of an image of a general member of the public. Furthermore, to infringe publicity rights, the use of image must be unauthorised, and so must be a use which is not consented to by the individual within the scope of mPort's Terms and Conditions. However, it is possible to argue that, while in the past perhaps only photographs of celebrities would have much in the way of commercial value, times have changed and data about non-famous individuals including images of their bodies also has commercial value,[44] especially when combined with other information about that individual to form a 'profile' which can then be used for commercial purposes such as targeted advertising. In addition, the extent to which certain uses of an individual's body scan image may be 'consented to' via an individual agreeing to mPort's expansive Terms and Conditions may be disputed, as the discussion above on data processing purposes in the context of privacy law explained. Yet, whether such expansive interpretations of publicity rights jurisprudence would be adopted for the 3D body scan scenario remains, in practice, an unanswered question.

---

[42] See: R. Zapparoni (2004) 'Propertising Identity: Understanding the United States Rights of Publicity and its Implications – Some Lessons for Australia' *Melbourne University Law Review*, 28, 690–723, 700.

[43] J. T. McCarthy (1995) 'The Human Persona as Commercial Property: The Right of Publicity' *Columbia-VLA Journal of Law & the Arts* 19, 129.

[44] See: S. Spickermann, A. Acquisti, R. Bohme and K. Hui (2015) 'The challenges of personal data markets and privacy' *Electronic Markets*, 25(2), 161.

CONCLUSION

As this chapter has explored, 3D scanning raises both existing and novel legal issues in the 3D printing space, for IP and data privacy given its real-world applications so far. Again, the practical impact of 3D scanning on law is of paramount importance. The relatively primitive nature of consumer-oriented 3D scanners suggests that, for the moment, 3D scanning does not present a significant threat to IP. However, the more sophisticated 3D scanners being used in retail to scan individuals' bodies for clothing and health reasons raise immediate privacy concerns over what happens to the very intimate data which is collected, for which existing privacy laws may seem inadequate. Inadequate privacy laws are not just a feature highlighted by this use of 3D scanning; many contemporary technological developments whereby 'privacy by design' is not embedded present significant threats to individuals' privacy as well as potentially enabling forms of discrimination based on the data collected about these individuals.[45]

---

[45] Peppet (2014) 'Regulating the Internet of Things'.

CHAPTER 5

Conclusion: Between Control and Chaos

Abstract 3D printing and 3D scanning bring both promises and risks for society and the laws currently in place, as the preceding chapters have explored. Here, the themes that have emerged in these chapters are drawn together and some thoughts about the overall topic of 3D printing's socio-legal aspects are given—to what extent the technology is susceptible to control, and to what extent it is creating 'chaos' by empowering decentralised individuals to create objects that previously they would be unable to do.

This book has looked at various areas of law which are affected by 3D printing. IP is perhaps the most prominent area discussed, with the whole of Chap. 2 dedicated to it, and some discussion as well in Chap. 4 on the IP implications of 3D scanning. Much initial discussion around 3D printing and the law was focused on 3D printing's interaction with IP, and comparisons made with what had previously happened with the Internet. In this book, this discussion has been explored, firstly by examining conceptually how IP law encounters 3D printing, and then by considering in practice what shape this interaction has taken.

Various points emerge from this. In principle, 3D printing opens up a Pandora's box of IP complexities (where and whether IP subsists; where and whether IP is infringed). In the Internet context, copyright was the area

95
A. Daly, *Socio-Legal Aspects of the 3D Printing Revolution*,
DOI 10.1057/978-1-137-51556-8_5

of IP mainly implicated by online file-sharing, and various legal and policy responses have emerged to address this—prominently the 'safe harbor' and takedown notice schemes for Internet intermediaries such as Internet service providers, social networks, and search engines which minimise their liability for their users' copyright infringements so long as they comply with orders to remove or block allegedly infringing user content. This can be contrasted with the lack of analogous intermediary liability schemes for other kinds of IP—patents, design rights, and trade marks—which 3D printing also involves. In addition, there are still various exceptions to infringement for these other areas of IP for personal, private, and/or non-commercial uses, under which much consumer or prosumer 3D printing activity may fall. It would seem that the difficulty in making objects which may infringe these other areas of IP—that is, the 'architectural' constraints[1]—prior to the advent of consumer-oriented 3D printing may have provided sufficient protection of patents, design rights, and trade marks, with little necessity to develop the intermediary liability and infringement exception regimes along similar lines to post-Internet copyright. However, the possibilities of reproducing objects protected by patents, design rights, and trade marks may provoke legislative and policy developments at the behest of the IP owners to extend stronger protection, develop an intermediary liability regime more in line with copyright, and remove exceptions which currently allow for personal, private, and/or non-commercial use.

Yet, it remains unclear precisely how much of a threat consumer- or prosumer-oriented 3D printing actually presents to incumbent industries' IP, despite the rhetoric around 'disruption'. Indeed, in practice, there has been a great deal of corporate interest in 3D printing and many from industry do not seem to see consumer use of 3D printers as constituting a major threat to their own IP. It is possible that this could change, particularly with the development of more powerful and user-friendly 3D printers for consumers and prosumers. Such a development may open up more possibilities for IP infringement. However, the interest from incumbent players in using 3D printing within their existing business models may well frustrate a bottom-up challenge from individuals to their IP rights, and in fact, it is individuals whose 3D printing design files uploaded to platforms such as Thingiverse may be more at risk from having their IP infringed by others misappropriating those designs or making vexatious claims of infringement.

---

[1] See: L. Lessig (1999) *Code and Other Laws of Cyberspace* (New York: Basic Books).

The complex interaction of 3D printing, existing laws, and actual practice is mirrored in other chapters. The laws around firearms are challenged by the emergence of the 3D printed Liberator. However, in reality, constructing guns using 3D printers is not a particularly practical option at this point in time, limiting the actual impact the Liberator currently has on this area of law. Yet the prospects raised by the Liberator and more effective and easy-to-use 3D printers do present challenges to the enforcement of gun laws, particularly in countries with much greater restrictions on making and using weapons than the USA. In any event, gun laws in existence in the USA and elsewhere are also premised on there being centralised points of control or gatekeepers which can be regulated, be it the manufacturer or dealer, and the concept of the self-production of guns goes some way to evading these regulatory nodes.

The inquiries in the rest of this book have been more theoretical than empirical, but have demonstrated the conceptual challenges 3D printing and scanning present to other legal regimes, which implicitly are based on the manufacture and distribution of products being carried out by centralised entities. For example, medical regulation has been based on the assumption that medical professionals are involved with a patient's access to, and use of, a medical device. 3D printed prosthetics created and distributed through networks such as e-NABLE do not require the intervention of a medical professional (although it is recommended). Product liability is another legal regime which is premised on the products at hand being produced and distributed by large companies in a commercial fashion, while 3D printing enables products to be produced at home in an amateur fashion. It may be that the objectives of these areas of law, to provide sufficient levels of safety in the creation and use of products, are not fulfilled by their application to consumer- or prosumer-oriented 3D printing.

However, what continues to happen in practice with 3D printing will be paramount to addressing the socio-legal aspects of the technology. It is certainly possible to imagine how more advanced machines becoming available at accessible prices would increase the instances of individuals engaging in 3D printing at home, and accordingly the threat to the effective enforcement of the laws examined in this book which more widespread use of 3D printing may cause. Examining practical developments in technology and its real-life applications is recommended as furthering socio-legal research and debate, and has relevance beyond 3D printing. Another emerging technological development in the form of cryptocurrencies, notably Bitcoin, and the blockchain technology, seems also to be infused with technoutopian discourses about the socially transformative possibilities—and the problems created for traditional forms of law enforcement and control—afforded by the technology,[2] while paying insufficient attention to what is actually happening in practice.

Yet another important practical phenomenon to watch is the extent to which the nation-state and large corporations are also using 3D printing. This will determine the 'disruptive' potential of 3D printing, since it can also be used as a technology of control by these actors as well as a technology to be used against their control by decentralised individuals. The discussion in this book has pointed to interest and current uses by both types of centralised actor in 3D printing. The extent to which 3D printing will evolve fully as a decentralised technology is thus far from certain due to the involvement of these large bureaucratic entities.

An important theme of this book is the relationship between the Internet and 3D printing. 3D printing is facilitated by the Internet, especially the distribution of, and access to, design files as input for the process. The experience with Internet regulation is illuminating for this discussion of socio-legal aspects of 3D printing, but only to a point. Already the business dynamics of the 3D printing ecosystem are diverging from those observed with the 1990s Internet. In particular, incumbents seem to acknowledge the benefits that 3D printing can bring to their businesses, whereas legacy content industries resisted digital innovations brought about by the Internet. This seems to be manifesting in an integration of 3D printing into existing industries, dulling its 'disruptive' effect. This can be seen by the consumer-oriented applications of 3D scanning discussed in the previous chapter, which are mediated by businesses. Rather than 'freeing' the individual from state and corporate control, the use of 3D body scanning booths actually gives rise to concerns about privacy and ('economic') surveillance[3] vis-à-vis the data that the scans generate and

---

[2] See: A. Wright and P. De Filippi (2015) 'Decentralized Blockchain Technology and the Rise of Lex Cryptographia', SSRN Working Paper, http://ssrn.com/abstract=2580664, accessed 3 November 2015.

[3] C. Fuchs (2011) 'Critique of the Political Economy of Web 2.0 Surveillance' in C. Fuchs, K. Boersma, A. Albrechtslund and M. Sandoval (eds.) *Internet and Surveillance: The Challenges of Web 2.0 and Social Media* (London: Routledge).

what might be done with it—sold on to other companies, or even accessed by the government.

In conclusion, time will tell how disruptive a technology 3D printing truly is in a socio-legal sense. However, given the political economy of 3D printing's development as a consumer-accessible technology, the involvement of the nation-state and large corporations as well as individuals in its use, it would seem that those who proclaimed 3D printing as a liberatory technology bringing about the end of scarcity and end of control—as with the Internet—have probably done so prematurely. This book has explored the theoretical aspects of 3D printing as a post-control and post-scarcity technology, and the theoretical complexities this creates for existing areas of law. However, this book has also considered the countervailing real-life forces which are shaping 3D printing's trajectory, and moving it within the influence of actors which are able to reimpose some measure of scarcity of objects and information and other forms of control over what end users do with it.

In this way, 3D printing may follow the Internet's trajectory as starting out as a seemingly uncontrollable and disruptive technology, but seeing the emergence of poles of power which are able to reimpose types of regulation. Indeed, as with the Internet, it may well be that the 'mainstream' experience of 3D printing is a safe and controlled one facilitated by the large 3D printer manufacturers' ecosystems, similar to what has happened with the use of 'closed' and less 'generative' devices to access the Internet, limiting user innovation but providing that safety and control.[4] It is true that there will still be some 'chaos around the edges' with determined individuals able to make their own 3D printers and access 3D printing files for undesirable objects if they know where and how to look. This is mirrored by the remaining 'ungovernable' (or difficult to govern) parts of the Internet at the edges with decentralised initiatives such as Tor, certain cryptocurrencies, and other activities 'under the radar' in the deep web.[5] Thus, there will not be a completely perfect enforcement of the laws vis-à-vis 3D

---

[4] Zittrain (2008) *The Future of the Internet and How to Stop It.*

[5] P. Biddle and others (2003) 'The Darknet and the Future of Content Protection' in E. Becker, W. Buhse, D. Gunnewig and N. Rump (eds.), *Digital Rights Management: Technological, Economic, Legal and Political Aspects* (Springer); P. De Filippi (2014) 'Bitcoin: a regulatory nightmare to libertarian dream' *Internet Policy Review* 3(2); L. J. Trautman (2014) 'Virtual Currencies: Bitcoin & What Now after Liberty Reserve, Silk Road, and Mt. Gox?' *Richmond Journal of Law and Technology* 20(4).

printing, as with the Internet, but as mentioned in the Introduction, this has already been the case even before these technological developments. It may well be more difficult to enforce laws in an increasingly decentralised society or economy, but the extent to which this technology-enabled decentralisation actually plays out is key.

Thus, those examining contemporaneous and future 'disruptive' technologies from a socio-legal perspective should be wary of making great technodeterministic proclamations without a careful examination of how these technologies are being used (or not) in the real world, and by whom.

BIBLIOGRAPHY

CASES

Australia
Peter Bodum A/S v DKSH Australia Pty Ltd (2011) IPR 222.

European Union
Case C-5/08 *Infopaq International A/S v Danske Dagblades Forening* [2010] ECR I-6569.
Case C-145/10 *Painer v Standard Verlags GmbH* [2012] ECDR 6 (ECJ (3rd Chamber)).

European Court of Human Rights
S and Marper v United Kingdom [2008] ECHR 1581.

United Kingdom
Antiquesportfolio.com plc v Rodney Fitch & Co Ltd [2001] FSR 23.
Autospin (Oil Seals) Ltd. v Beehive Spinning (A Firm) [1995] RPC 683.
BBC Worldwide and Anor v Pally Screen Printing and others [1988] FSR 665.
Caparo Industries plc v Dickman [1990] 2 AC 605, HL.
Francis Day Hunter v Bron [1963] Ch. 587.
Lucasfilm Ltd v Ainsworth [2011] UKSC 39.
Mackie Designs v Behringer Specialised Studio Equipment and others [1999] RPC 717.
Rotocrop v Genbourne [1982] FSR 241.

© The Editor(s) (if applicable) and The Author(s) 2016
A. Daly, *Socio-Legal Aspects of the 3D Printing Revolution*,
DOI 10.1057/978-1-137-51556-8

United Wire Ltd v Screen Repair Services (Scotland) Ltd. [2001] FSR 24.

United States of America
Bernstein v. U.S. Dep't of State (Bernstein I), 922 F. Supp. 1426 (N.D. Cal. 1996).
Bernstein v. United States Dep't of State, 945 F. Supp. 1279 (N.D. Cal.1996).
Bernstein v United States (1999) Case Number: 97-16686 (9th Circuit Court of Appeal).
Bridgeman Art Library v Corel Corporation, 25 F.Supp 2d 421 (S.D.N.Y. 1987), modified 36 F.Supp.2d 191 (S.D.N.Y. 1999).
Bright Tunes Music v. Harrisongs Music 420 F. Supp. 177 (S.D.N.Y. 1976).
Defense Distributed and Second Amendment Foundation v US Department of State Case No 1:15-cv-372.
District of Columbia v Heller (2008) No 07-290.
Hansen v. Baxter Healthcare Corp., 764 N.E.2d 35, 42 (Ill. 2002).
Husky Injection Moulding System Ltd v R&D Tool and Engineering Company, 291 F.3d 780 (Fed Cir 2002).
Junger v. Daley, 8 F. Supp. 2d 708 (N.D. Ohio 1998).
Junger v. Daley, 209 F.3d 481 (6th Cir. 2000).
Meshwerks v Toyota Motor Sales, 528 F.3d 1258 (10th Cir. 2008).
New York Times v United States, 403 US 713.
Qualitex Co v Jacobson Products Co, 514 US 159, 162 (1995).
Reno v American Civil Liberties Union 521 U.S. 844 (1997).
Sorrell v IMS Health 131 S.Ct. 2653 (2011).
Traffix Devices v Marketing Displays, 532 US 23, (2001).
Wal-Mart Stores v Samara Brothers, 529 US 205.

LEGISLATION

International
Anti-Counterfeiting Trade Agreement (adopted 1 October 2011, not yet entered into force) ('ACTA').
Agreement on Trade-Related Aspects of Intellectual Property Rights (entry into force 1 January 1995) ('TRIPs Agreement').
Berne Convention for the Protection of Literary and Artistic Works (adopted 9 September 1886, entered into force 5 December 1887) ('Berne Convention').
Convention for the Protection of Human Rights and Fundamental Freedoms (adopted 4 November 1950, entered into force 3 September 1953) ETS 5 (European Convention on Human Rights, as amended) ('ECHR').
Paris Convention for the Protection of Industrial Property (adopted 20 March 1883) ('Paris Convention').

Trans Atlantic Trade and Investment Partnership (currently under negotiation) ('TTIP').

Trans Pacific Partnership (currently under negotiation) ('TPP').

European Union

Regulations

Council Regulation (EC) 6/2002 of 12 December 2001 on Community Designs [2002] OJ L3/1.

Council Regulation (EC) 764/2008 of 9 July 2008 laying down procedures relating to the application of certain national technical rules to products lawfully marketed in another Member State and repealing Decision No 3052/95/EC [2008] OJ L218/21.

Council Regulation (EC) 765/2008 of 9 July 2008 setting out the requirements for accreditation and market surveillance relating to the marketing of products and repealing Regulation (EEC) No 339/93 [2008] OJ L218/30.

Proposal for a Regulation of the European Parliament and of the Council on *in vitro* diagnostic medical devices, and amending Directive 2001/83/EC, Regulation (EC) No 178/2002 and Regulation (EC) No 1223/2009, COM(2012) 542 final.

Directives

Council Directive 85/374/EEC of 25 July 1985 on the approximation of the laws, regulations and administrative provisions of the Member States concerning liability for defective products [1985] OJ L210/29.

Council Directive 93/42/EEC concerning medical devices [1993] OJ L169/1 ('Medical Devices Directive').

Council Directive 95/46/EC of 24 October 1995 on the protection of individuals with regard to the processing of personal data and on the free movement of such data [1995] OJ L 281/31 ('Data Protection Directive').

Council Directive 98/71/EC of 13 October 1998 on the legal protection of designs [1998] OJ L289/28.

Council Directive 1999/34/EC of 10 May 1999 amending Council Directive 85/374/EEC on the approximation of the laws, regulations and administrative provisions of the Member States concerning liability for defective products [1999] OJ L141/20.

Council Directive 2009/24/EC of 23 April 2009 on the legal protection of computer programs (codified version) [2009] OJ L111/16.

Decisions

Council Decision (EC) 768/2008/EC of 9 July 2008 on a common framework for the marketing of products, and repealing Council Decision 93/465/EEC [2008] OJ L218/82.

United Kingdom
Copyright, Designs and Patents Act 1988.
Patents Act 1977.
Registered Designs Act 1949.
Trade Marks Act 1995.

United States of America
17 U.S.C. § 101–02 (2006).
Constitution of United States of America 1789 (revised 1992).
International Traffic in Arms Regulations, 22 C.F.R. § 120.17(a)(4) (1996).
Restatement (Third) of Torts: Product Liability (1998).

Australia
Privacy Act (Cth) 1989.
Weapons (Digital 3D and Printed Firearms) Amendment Bill 2014 (QLD).

SECONDARY SOURCES

3ders. (2014). Authentise launches streaming service for 3D print files. http://www.3ders.org/articles/20140404-authentise-launches-streaming-service-for-3d-print-files.html. Accessed 11 Sept 2015.

3D Printing Industry. (2014). History of 3D printing. http://3dprintingindustry.com/3d-printing-basics-free-beginners-guide/history/. Accessed 10 Sept 2015.

Adams, M. (2013). The 'Third Industrial Revolution': 3D printing technology and Australian designs law. Bachelor of Laws (Honours) thesis, Monash University.

Andersen, J., & J. Howells, J. (2014). The intellectual property rights implications of consumer 3D printing. Thesis, Aarhus University Department of Business Administration School of Business and Social Sciences.

Anderson, C. (2012). Makers: The new industrial revolution. New York: Crown Business.

Arizmendi, C., Pronk, B., & Choi, J. (2014) Services no longer required? Challenges to the states as the primary security provider in the age of digital fabrication. Small Wars Journal.

Article 29 Working Party. Opinion 02/2012 on facial recognition in online and mobile services (WP192).

Article 29 Working Party. Opinion 03/2012 on developments in biometric technologies (WP193).

Ashcraft, B. (2014). Japanese man arrested for having guns made with a 3D printer. Kotaku, http://kotaku.com/japanese-man-arrested-for-having-guns-made-with-a-3d-pr-1573358490. Accessed 11 Sept 2015.

Australian Senate Legal and Constitutional Affairs References Committee. (2015). Ability of Australian law enforcement authorities to eliminate gun-related violence in the community.

Ballardini, R. M., Norrgård, M., & Minssen, T. (2015). Enforcing patents in the era of 3D printing. *Journal of Intellectual Property and Practice, 10*(11), 850–866.

Barlow, J. P. (1996). A declaration of the independence of cyberspace. https://projects.eff.org/~barlow/Declaration-Final.html. Accessed 10 Sept 2015.

Benkler, Y. (2000). From consumers to users: shifting the deeper structures of regulations towards sustainable commons and user access. *Federal Communications Law Journal, 52*, 561–579.

Benkler, Y. (2006). *The wealth of networks: How social production transforms markets and freedom.* New Haven: Yale University Press.

Berkowitz, N. D. (2015). Strict liability for individuals? The impact of 3-D printing on products liability law. *Washington University Law Review, 92*(4), 1019–1053.

Better Future Factory. (2012). Perpetual plastic project. http://www.betterfuturefactory.com/work/perpetual-plastic-project-ppp. Accessed 10 Sept 2015.

Birnhack, M. D., & N. Elkin-Koren, N. (2003). The invisible handshake: The reemergence of the state in the digital environment. *Virginia Journal of Law & Technology, 8*(6), 1–57.

Birtchnell, T., & Hoyle, W. (2014). *3D printing for development in the global south.* Basingstoke/New York: Palgrave Macmillan.

Blackman, J. (2014). The 1st amendment, 2nd amendment, and 3D printed guns. *Tennessee Law Review, 81*, 479–538.

Boldrin, M., & Levine, D. K. (2008). *Against intellectual monopoly.* Cambridge: Cambridge University Press.

Bowyer, A. (2006). Keynote address on the RepRap project. Seventh national conference on rapid design, prototyping & manufacturing, high Wycombe. http://reprap.org/wiki/PhilosophyPage. Accessed 10 Sept 2015.

Boyle, J. (2003). The second enclosure movement and the construction of the public domain. *Law and Contemporary Problems, 66*, 33–74.

Bradshaw, S., Bowyer, A., & Haufe, P. (2010). The intellectual property implications of low-cost 3D printing. *SCRIPTed, 7*(1), 5–31.

Brean, D. H. (2013). Asserting patents to combat infringement via 3D printing: It's no "use". *Fordham Intellectual Property, Media & Entertainment Law Journal, 23*, 771.

Bunn, A. (2013). Facebook and face recognition: kinda cool, kinda creepy. *Bond Law Review, 25*(1), 35–69.

Clarke, R. (2001). Biometrics and privacy. http://www.rogerclarke.com/DV/Biometrics.html. Accessed 12 Sept 2015.

Cohen, J. (2012). *Configuring the networked self: Law, code and everyday practice.* New Haven: Yale University Press.

Coleman, E. G. (2012). *Coding freedom: The ethics and aesthetics of hacking.* Princeton: Princeton University Press.

Cosans, J. (2014). Between firearm regulation and information censorship: Analyzing first amendment concerns facing the world's first 3-D printed plastic gun. *Journal of Gender, Social Policy and Law, 22*(4), 915–946.

Courtland, R. (2013). Resources profile: Bre Pettis. *IEEE Spectrum.* http://spectrum.ieee.org/geek-life/profiles/bre-pettis. Accessed: 10 Sept 2015.

Cowan, P. (2015). Qld Govt knocks back 3D-printed guns bill. *IT News.* http://www.itnews.com.au/News/403827,qld-govt-knocks-back-3d-printed-guns-bill.aspx. Accessed 12 Sept 2015.

Cumptson, B., Lipson, M., Marder, S. R., & Perry, J. W. (1999). Two-photon or higher-order absorbing optical materials. *US Patent Application PCT/US19991008383.* http://www.google.com/patents/WO1999053242A1?cl=en. Date accessed 10 Sept 2015.

Daly, A. (2015a). *Mind the gap: Private power, online information flows and EU law.* PhD thesis, European University Institute.

Daly, A. (2015b). The law and ethics of 'self-quantified' health information: an Australian perspective. *International Data Privacy Law, 5*(2), 144–155.

Daly, A., & Farrand, B. (2015). SABAM v Scarlet: evidence of an emerging backlash against corporate copyright lobbies in Europe?'. In D. DeVoss & M. Rife (Eds.), *Cultures of Copyright.* New York: Peter Lang.

D'Aveni, R. (2015). The 3-D printing revolution. *Harvard Business Review.* https://hbr.org/2015/05/the-3-d-printing-revolution. Accessed 10 Sept 2015.

De Filippi, P., & McCarthy, S. (2012). Cloud computing, centralization and data sovereignty. *European Journal of Law and Technology, 3*(2).

De Souza e Silva, A., & Frith, J. (2010). Locational privacy and public spaces: Media discourses on location-aware mobile technologies. *Communication, Culture & Critique, 3*(4), 503–525.

DeLong, J. B., & Froomkin, A. M. (2000). Speculative microeconomics for tomorrow's economy. *First Monday, 5*(2).

Desai, D., & Magliocca, G. (2013–2014). Patents, meet napster: 3D printing and the digitization of things. *Georgetown Law Journal, 102,* 1691.

Doctorow, C. (2012). Congressman calls for ban on 3D printed guns. *Boing Boing.* http://boingboing.net/2012/12/09/congressman-calls-for-ban-on-3.html. Accessed 14 June 2015.

Dodson, S. (2008). The machine that copies itself. *The Guardian.* http://www.theguardian.com/technology/2008/jul/03/copy.machine.reprap. Accessed 2 Sept 2015.

Ducklin, P. (2013). How effective are data breach penalties? Are ever-bigger fines enough?. *Nakedsecurity.* http://nakedsecurity.sophos.com/2013/04/26/how-effective-are-data-protection-regulations/. Accessed 12 Sept 2015.

Dyson, E., Gilder, G., Keyworth, G., & Toffler, A. (1994). Cyberspace and the American dream: A Magna Carta for the knowledge age. http://www.pff.org/issues-pubs/futureinsights/fi1.2magnacarta.html. Accessed 10 Sept 2015.

Enabling the Future. http://enablingthefuture.org/. Accessed 12 Sept 2015.

Enabling the Future. *FAQs (General).* http://enablingthefuture.org/faqs-general/. Accessed 12 Sept 2015.

Enabling the Future. *Media FAQ.* http://enablingthefuture.org/faqs/media-faq/. Accessed 12 Sept 2015.

Enabling the Future. *Safety Guidelines.* http://enablingthefuture.org/build-a-hand/safety-guidelines/. Accessed 12 Sept 2015.

Enabling the Future. *Upper limb prosthetics.* http://enablingthefuture.org/upper-limb-prosthetics/. Accessed 12 Sept 2015.

Enoch, D. (2014). Tort Liability and Taking Responsibility'. In J. Oberdiek (Ed.), *Philosophical foundations of the law of torts.* Oxford: Oxford University Press.

ETH Zurich. Transform your smartphone into a mobile 3D scanner. http://www.inf.ethz.ch/news-and-events/spotlights/mobile_3dscanner.html. Accessed 10 Sept 2015.

EUROPOL. (2014). 31 arrests in operation against Bulgarian Organised Crime Network. https://www.europol.europa.eu/content/31-arrests-operation-against-bulgarian-organised-crime-network. Accessed 11 Sept 2015.

Farviar, C. (2013). 3D-printed gun maker now has federal firearms license to manufacture, deal guns. *Arstechnica.* http://arstechnica.com/tech-policy/2013/03/3d-printed-gunmaker-now-has-federal-firearms-license-to-manufacture-deal-guns/. Accessed 11 Sept 2015.

Feige, E. L. (Ed.). (2007). *The underground economies: Tax evasion and information distortion.* Cambridge: Cambridge University Press.

Ferenstein, G. (2013). Offshore 3D printed gun blueprint protector kim dotcom reportedly deleting files. *TechCrunch.* http://techcrunch.com/2013/05/11/offshore-3d-printed-gun-blueprint-protector-kim-dotcom-reportedly-deleting-files/. Accessed 11 Sept 2015.

Flaherty, D. H. (1988). The emergence of surveillance societies in the western world: Toward the year 2000. *Government Information Quarterly,* 5(4), 377.

Fordyce, R. (2015). Manufacturing imaginaries: Neo-Nazis, men's rights activists and 3D printing. *Journal of Peer Production,* (6). Disruption and the Law.

Frank-Jackson, D. (2011). The medical device federal preemption trilogy: Salvaging due process for injured plaintiffs. *Southern Illinois University Law Journal,* 35, 453–497.

Freeman Engstrom, N. (2013). 3-D printing and product liability: Identifying the obstacles. *University of Pennsylvania Law Review Online,* 162(35), 35–41.

Fuchs, C. (2011). Critique of the political economy of web 2.0 surveillance. In C. Fuchs, K. Boersma, A. Albrechtslund, & M. Sandoval (Eds.), *Internet and surveillance: The challenges of web 2.0 and social media.* London: Routledge.

Goldsmith, J. L., & Wu, T. (2006). *Who controls the internet? Illusions of a borderless world.* Oxford: Oxford University Press.

Goyanes, A., Robles Martinez, P. Buanz, A., Basit, A. W., & Gaisford, S. (2015). Effect of geometry on drug release from 3D printed tablets. *International Journal of Pharmaceutics,* 494(2), 657-663.

Graef, I., Verschakelen, J., & Valcke, P. (2013). Putting the right to data portability into a competition law perspective. *Law: The Journal of the Higher School of Economics Annual Review, 53–63.*

Grant, L. (2014). Bits to bullets: Australian military 3DP's new war-making strategies and tactics' 3D printing industry. http://3dprintingindustry.com/2014/07/14/bits-bullets-australian-military-3dps-new-war-making-strategies-tactics/. Accessed 5 Sept 2015.

Greatorex, G. (2015). 3D printing and consumer product safety. Product safety solutions white paper.

Greenberg, A. (2012). "Wiki Weapon Project" aims to create a gun anyone can 3D-print at home. *Forbes.* http://www.forbes.com/sites/andygreenberg/2012/08/23/wiki-weapon-project-aims-to-create-a-gun-anyone-can-3d-print-at-home/. Accessed 11 Sept 2015.

Greenberg, A. (2013a). Meet the 'Liberator': Test-firing the world's first fully 3D-printed gun. *Forbes.* http://www.forbes.com/sites/andygreenberg/2013/05/05/meet-the-liberator-test-firing-the-worlds-first-fully-3d-printed-gun/. Accessed 11 Sept 2015.

Greenberg, A. (2013b). 3D-printed gun's blueprints downloaded 100,000 times in two days (with some help from Kim Dotcom). *Forbes.* http://www.forbes.com/sites/andygreenberg/2013/05/08/3d-printed-guns-blueprints-downloaded-100000-times-in-two-days-with-some-help-from-kim-dotcom/. Accessed 11 Sept 2015.

Greenberg, A. (2013c). State department demands takedown of 3D-printable gun files for possible export control violations. *Forbes.* http://www.forbes.com/sites/andygreenberg/2013/05/09/state-department-demands-takedown-of-3d-printable-gun-for-possible-export-control-violation/. Accessed 11 Sept 2015.

Greenberg, A. (2015a). 3-D printed gun lawsuit starts the war between arms control and free speech. *Wired.* http://www.wired.com/2015/05/3-d-printed-gun-lawsuit-starts-war-arms-control-free-speech/. Accessed 11 Sept 2015.

Greenberg, A. (2015b). Bill to ban undetectable 3D printed guns is coming back. *Wired.* http://www.wired.com/2015/04/bill-ban-undetectable-3-d-printed-guns-coming-back/accessed. 12 Sept 2015.

Greenleaf, G. (2008). Privacy in Australia'. In J. B. Rule & G. Greenleaf (Eds.), *Global Privacy Protection: The First Generation.* Cheltenham: Edward Elgar.

Gridneff, I. (2013). 3D-printed gun 'will kill', police warn. *Sydney Morning Herald.* http://www.smh.com.au/digital-life/digital-life-news/3dprinted-gun-will-kill-police-warn-20130524-2k59g.html. Accessed 12 Sept 2015.

Howell, B. (2004). Medical misadventure and accident compensation in New Zealand: An incentives-based analysis. *Victoria University of Wellington Law Review, 35,* 857–878.

Information Age. (2011). Facebook facial recognition breaks EU law – regulator. http://www.information-age.com/technology/security/1669438/facebook-facial-recognition-breaks-eu-law---regulator. Accessed 12 Sept 2015.

Israel, S. (2013). Rep. Israel introduces bipartisan undetectable firearms modernization act to protect Americans from threat of plastic guns. *Press release.*

http://israel.house.gov/media-center/press-releases/rep-israel-introduces-bipartisan-undetectable-firearms-modernization-act. Accessed 14 Sept 2015.

Jensen-Haxel, P. (2012). 3D printers, obsolete firearm supply controls, and the right to build self-defense weapons under Heller. *Golden Gate University Law Review, 42*(3), 447–495.

Kahler, A. (2013). I got a DMCA takedown notice from Makerbot/thingiverse for this. *Google+*. https://plus.google.com/112825668630459893851/posts/e7sZ8Gw6umx. Accessed 11 Sept 2015.

Kayser, M. (2011). The solar sinter. http://www.dezeen.com/2011/06/28/the-solar-sinter-by-markus-kayser/. Accessed 10 Sept 2015.

Khaled, S. A., Burley, J. C., Alexander, M. R., & Roberts, C. J. (2014). Desktop 3D printing of controlled release pharmaceutical bilayer tablets. *International Journal of Pharmaceutics, 461*, 105–111.

Krassenstein, B. (2014). Two year sentence handed down to Yoshitomo Imura in Japanese 3D printed gun case. *3DPrint.com*. http://3dprint.com/20019/sentence-imura-3d-printed-gun/. Accessed 11 Sept 2015.

Langvardt, K. (2014). The replicator and the first amendment. *Fordham Intellectual Property Media Entertainment and Law Journal, 25*(1), 59–115.

Leach, A. (2014). 3D printed prosthetics: long-term hope for amputees in Sudan. *The Guardian*. http://www.theguardian.com/global-development-professionals-network/2014/jun/13/3d-printing-south-sudan-limbs. Accessed 12 Sept 2015.

Leber, J. (2013). A DIY bioprinter is born. *MIT Technology Review*. http://www.technologyreview.com/view/511436/a-diy-bioprinter-is-born/. Accessed 12 Sept 2015.

Lemley, M. (2014). 'IP in a World Without Scarcity' Stanford Public Law Working Paper No. 2413974.

Lessig, L. (1999). *Code and other laws of cyberspace*. New York: Basic Books.

Li, P., Mellor, S., Griffin, J., Waelde, C., Hao, L., & Everson, R. (2014). Intellectual property and 3D printing: a case study on 3D chocolate printing. *Journal of Intellectual Property Law & Practice, 9*(4), 322–332.

Lobato, R. (2012). *Shadow economies of cinema: Mapping informal film distribution*. Basingstoke/New York: Palgrave Macmillan.

Lobmayr, B. (2010). An assessment of the EU approach to medical device regulation against the backdrop of the US system. *European Journal of Risk Regulation, 1*(2), 137–149.

Lu, Y. J. (2010). The change in knowledge proposal: Repairing preemption doctrine in medical products liability. *SSRN Working Paper*. http://ssrn.com/abstract=1957954. Accessed 12 Sept 2015.

Lyon, D. (2014). Surveillance, snowden and big data: Capacities, consequences, critique. *Big Data and Society*, 1–13.

Mac Sithigh, D. (2013). App law within: rights and regulation in the smartphone age. *International Journal of Law and Information Technology, 21*(2), 154–186.

Maly, T. (2012). Thingiverse remotes (most) printable gun parts. *Wired*. http://www.wired.com/2012/12/thingiverse-removes-gun-parts/. Accessed 11 Sept 2015.

Margoni, T. (2013). Not for designers: On the inadequacies of EU design law and how to fix it. *Journal of Intellectual Property, Information Technology and E-Commerce Law, 4*(3), 225–248.

Martinez, F. (2012). Indiegogo shuts down campaign to develop the world's first printable gun. *Daily Dot.* https://www.dailydot.com/news/indiegogo-3d-printed-gun-campaign/. Accessed 11 Sept 2015.

Marx, G. T. (1985). The surveillance society. *The Futurist, 19*(3), 21–26.

Mazerolle, L., Soole, D., & Rombouts, S. (2007). Drugs law enforcement: A review of the evaluation literature. *Police Quarterly, 10*(2), 115–153.

McCarthy, J. T. (1996). The human persona as commercial property: The right of publicity. *Columbia-VLA Journal of Law & the Arts, 19,* 129.

Mendis, D. (2013). "The clone wars" – Episode 1: The rise of 3D printing and its implications for intellectual property law – Learning lessons from the past? *European Intellectual Property Review, 35*(5), 155–169.

Mendis, D. (2014). "Clone wars": Episode II – The next generation: The copyright implications related to 3D printing and computer-aided design (CAD) files. *Law, Innovation and Technology, 6*(2), 265–281.

Mendis, D. and Secchi, D. (2015). *A legal and empirical study of 3D printing online platforms and an analysis of user behaviour.* Study I, UK Intellectual Property Office.

Meese, J. (2014). Google Glass and Australian Privacy Law: Regulating the Future of Locative Media'. In R. Wilken & G. Goggin (Eds.), *Locative Media'.* London: Routledge.

Michael, G. J. (2013). Anarchy and property rights in the virtual world: How disruptive technologies undermine the state and ensure that the virtual world remains a 'wild west'. *SSRN Working Paper.* http://papers.ssrn.com/sol3/papers.cfm?abstract_id=2233374. Accessed 10 Sept 2015.

Michael, M. (2014). Process and Plasticity: Printing, Prototyping and the Prospects of Plastic'. In J. Gabrys, G. Hawkins, & M. Michael (Eds.), *Accumulation: The Material Politics of Plastic.* London: Routledge.

Moilanen, J., Daly, A., Lobato, R., & Allen, D. (2015). Cultures of sharing in 3D printing: what can we learn from the licence choices of thingiverse users?' *Journal of Peer Production,* (6). Disruption and the Law.

Molich-Hou, M. (2015). Rep. Steve Israel renews fight for undetectable gun control. *3D Printing Industry.* http://3dprintingindustry.com/2015/06/10/rep-steve-israel-renews-fight-for-undetectable-gun-control/. Accessed 12 Sept 2015.

mPort. *FAQ.* http://mport.com.au/home/faq. Accessed 12 Sept 2015.

mPort. FAQ. *Privacy policy.* http://mport.com.au/home/privacypolicy. Accessed 12 Sept 2015.

Murphy, S. V., & Atala, A. (2014). 3D bioprinting of tissues and organs. *Nature Biotechnology, 32,* 773–785.

Nielson, H. (2015). Manufacturing consumer protection for 3-D printed products. *Arizona Law Review, 57*(2), 609–622.

Nolan, D., & Davies, J. (2013). Torts and Equitable Wrongs'. In A. Burrows (Ed.), *English Private Law*. Oxford: Oxford University Press.

Nguyen, T. (1997). Cryptography, export controls, and the first amendment. *Bernstein v. United States Department of State*' *Harvard Journal of Law & Technology, 10*(3), 667–682.

Oliphant, K. (2008). Accident Compensation in New Zealand: An Overview'. In G. Schamps (Ed.), *Evolution des droits du patient, indemnisation sans faute des dommages lies aux soins de sante : le droit medical en movement*. Editions Bruylant: Brussels.

Osborn, L. (2014a). Of PhDs, pirates, and the public. *Texas A&M Law Review, 1,* 811–835.

Osborn, L. (2014b). Regulating three-dimensional printing: The converging worlds of bits and atoms. *San Diego Law Review, 51,* 553–621.

Ouyang, B. (2014). 3D printing low-cost prosthetics parts in Uganda. *Med Gadget.* http://www.medgadget.com/2014/03/3d-printing-low-cost-prosthetics-parts-in-uganda.html. Accessed 12 Sept 2015.

Paikin, S. 3D printing: A killer app. The Agenda. https://www.youtube.com/watch?v=eN_cVRjIrwg. Accessed 11 Sept 2015.

Park, M. H. (2015). For a new heart, just click print: The effect on medical and products liability From 3-D printed organs. *Journal of Law, Technology and Policy, 1,* 187–210.

Pasquale, F. (2014). 'Symbiotic Law & Social Science: The Case for Political Economy in the Legal Academy, and Legal Scholarship in Political Economy' (Jotwell 5th Anniversary Conference, Miami).

Pasquale, F. (2015). *The black box society: The secret algorithms that control money and information*. Cambridge: Harvard University Press.

Pedersen, J. M. (2010). Conclusion: Property and the politics of commoning. *The Commoner, 14,* 287–294.

Pearce, R. (2013). NSW police issues warning on 3D printed guns. *Computer World.* http://www.computerworld.com.au/article/462774/nsw_police_issues_warning_3d_printed_guns/. Accessed 12 Sept 2015.

Pearce, J. M. (2015). Applications of open source 3-D printing on small farms. *Organic Farming, 1*(1), 19–35.

Pearce, J. M., & Hasselhuhn, A. S. (2015). Intellectual property as a strategic national industrial weapon: the case of 3D printing. *Engineer: The Professional Bulletin of Army Engineers, 45*(2), 29–31.

Peppet, S. R. (2014). Regulating the internet of things: First steps toward managing discrimination, privacy, security & consent. *Texas Law Review, 93,* 85–176.

Portes, A., Castells, M., & Benton, L. A. (1989). *The informal economy: Studies in advanced and less developed countries*. Baltimore: John Hopkins University Press.

Post, R. (2000). Encryption source code and the first amendment. *Berkeley Technology Law Journal, 15,* 713–724.

Record, I., coons, g., Southwick, D., & Ratto, M. (2015). Regulating the liberator: Prospects for the regulation of 3D printing. *Journal of Peer Production*, (6). Disruption and the Law.

Reeves, P.. & Mendis, D. (2015). *The current status and impact of 3D printing within the industrial sector: An analysis of six case studies*. Study II, UK Intellectual Property Office.

Reidenberg, J. (2005). Technology and internet jurisdiction. *University of Pennsylvania Law Review, 153*(6), 1951–1974.

RepRap. (2014). About. http://reprap.org/. Accessed 10 Sept 2015.

Resnik, L., Klinger, S., Krauthamer, V., & Barnabe, K. (2010). U.S. food and drug administration regulation of prosthetic research, development, and testing. *Journal of Prosthetics and Orthotics, 22*(2), 121–126.

Rideout, B. (2011). Printing the impossible triangle: The copyright implications of three-dimensional printing. *Journal of Business Entrepreneurship and the Law, 5*(1), 160–177.

Rifkin, J. (2014). *The zero marginal cost society: The internet of things, the collaborative commons, and the eclipse of capitalism*. Basingstoke/New York: Palgrave Macmillan.

Roberts, D. (2013). 3D-printed guns prompt US House to renew prohibition on plastic firearms'. *The Guardian*. http://www.theguardian.com/world/2013/dec/04/3d-guns-house-renew-prohibition-plastic-firearms. Accessed 12 Sept 2015.

Samuelson, P. (1988). American software copyright law. *Columbia-VLA Journal of Law & the Arts, 13*, 61–75.

Samuelson, P., & Scotchmer, S. (2002). The law and economics of reverse engineering. *Yale Law Journal, 111*(70), 1575–1664.

Scardamaglia, A. (2012). Protecting product shapes and features: beyond designs and trade marks in Australia'. *Journal of Intellectual Property Law & Practice, 7*(3), 159–161.

Scardamaglia, A. (2014). Keywords, Trademarks and Search Engine Liability'. In R. Konig & M. Rasch (Eds.), *Society of the Query Reader: Reflections on Web Search*. Amsterdam: Institute for Network Cultures.

Scardamaglia, A. (2015). Flashpoints in 3D printing and trade mark law. *Journal of Law, Information & Science*. (forthcoming).

Seng, D. (2014). The state of the discordant union: An empirical analysis of DMCA takedown notices. *Virginia Journal of Law and Technology, 18*(3), 370–473.

Shapeways. (2015). Terms and conditions. http://www.shapeways.com/terms_and_conditions. Accessed 12 Sept 2015.

Sher, D. (2014). Kenya based 3D life print project is offering mobile 3D printing of custom prosthetics. *3D printing industry*. http://3dprintingindustry.com/2014/12/08/3d-life-print-3d-printing-prosthetics/. Accessed 12 Sept 2015.

Simon, M. (2013). When copyright can kill: How 3D printers are breaking the barriers between "intellectual" property and the physical world. *Pace Intellectual Property Sports and Entertainment Law Forum, 3*(1), 59–97.

Söderberg, J., & Daoud, A. (2012). Atoms want to be free too! Expanding the critique of intellectual property to physical coods. *Triple C Communications Capitalism & Critique, 10*(1).

Sosa, F. (2015a). Left Shark. *Thingiverse*. http://www.thingiverse.com/thing:667127. Accessed 11 Sept 2015.

Sosa, F. (2015b). Politicalsculptor retains legal representation and responds to Katy Perry's Law Firm. http://politicalsculptor.blogspot.com.au/2015/02/politicalsculptor-retains-legal.html. Accessed 11 Sept 2015.

Sosa, F. (2015c). Prior art claim. http://politicalsculptor.blogspot.com.au/2015/02/prior-art-claim.html. Accessed 11 Sept 2015.

Sosa, F. (2015d). Katy Perry Law Firm responds and so does Political Sculptor. http://politicalsculptor.blogspot.com.au/2015/02/katy-perry-law-firm-responds-and-so.html. Accessed 11 Sept 2015.

Spickermann, S., Acquisti, A., Bohme, R., & Hui, K. (2015). The challenges of personal data markets and privacy *Electronic Markets, 25*(2), 161.

Stapleton, J. (2000). Restatement (third) of torts: products liability, an Anglo-Australian perspective. *Washburn Law Journal, 39*, 363–403.

Streams, K. (2012). 3D printed gun project halts after Stratasys confiscates rented printer. *The Verge*. http://www.theverge.com/2012/10/1/3439496/wiki-weapon-project-defense-distributed-stratasys. Accessed 11 Sept 2015.

Svantesson, D. (2014). Sovereignty in international law – how the internet (maybe) changed everything, but not for long'. *Masaryk University Journal of Law and Technology, 8*(1), 137–155.

Svantesson, D., & Clarke, R. (2010). Privacy and consumer risks in cloud computing. *Computer Law & Security Review, 26*(4), 391–397.

Tien, L. (2000). Publishing software as a speech act. *Berkeley Technology Law Journal, 15*, 629–712.

The Economist. (2012). The third industrial revolution. http://www.economist.com/node/21553017. Accessed 10 Sept 2015.

Thompson, C. (2012). 3D printing's forthcoming legal morass. *Wired*. http://www.wired.co.uk/news/archive/2012-05/31/3d-printing-copyright. Accessed 11 Sept 2015.

Tran, J. L. (2015). To bioprint or not to bioprint. *North Carolina Journal of Law and Technology, 17*, forthcoming.

Trimensional. http://www.trimensional.com/. Accessed 10 Sept 2015.

Twining, W. (2008). Law in Context movement'. In P. Cane & J. Conaghan (Eds.), *The New Oxford Companion to Law*. Oxford: Oxford University Press.

U.S. Copyright Office. (2015). Copyright registration for works of the visual arts. *Circular 40*. www.copyright.gov/circs/circ40.pdf. Accessed 11 Sept 2015.

U.S. Department of Justice Bureau of Alcohol, Tobacco, Firearms and Explosives. (2015). What is ATF doing in regards to people making their own firearms? https://www.atf.gov/firearms/qa/what-atf-doing-regards-people-making-their-own-firearms. Accessed 11 Sept 2015.

U.S. Food and Drug Administration. *General controls for medical devices.* http://www.fda.gov/MedicalDevices/DeviceRegulationandGuidance/Overview/GeneralandSpecialControls/ucm055910.htm#QSR. Accessed 12 Sept 2015.

Varley, R., & Eaton, M. (2015). 3D printing: Suspected plastic gun parts found in raid on Gold Coast property. *ABC.* http://www.abc.net.au/news/2015-02-10/3d-printing-police-suspect-plastic-parts-belong-to-homemade-gun/6083938. Accessed 12 Sept 2015.

Vollebregt, E. (2014). 3D printing of custom medical devices under future EU law. *Medical Devices Legal.* http://medicaldeviceslegal.com/2014/03/05/3d-printing-of-custom-medical-devices-under-future-eu-law/. Accessed 12 Sept 2015.

Volsky, I. (2013). Philadelphia becomes first city to ban 3D guns. *Think Progress.* http://thinkprogress.org/justice/2013/11/23/2987911/philadelphia-city-ban-guns/. Accessed 12 Sept 2015.

Ward, M. (2014). Tor's most visited hidden sites host child abuse images. *BBC News.* http://www.bbc.com/news/technology-30637010. Accessed 11 Sept 2015.

Weinberg, M. (2010). *It will be awesome if they don't screw it up: 3D printing, intellectual property, and the fight over the next great disruptive technology.* Public Knowledge White Paper.

Weinberg, M. (2013). *What's the deal with copyright and 3D printing.* Public Knowledge White Paper.

Weinberg, M. (2015). *3D printed copyright creep.* Techdirt. https://www.techdirt.com/articles/20150427/10532430809/3d-printed-copyright-creep.shtml. Accessed 11 Sept 2015.

Wells, J. C. K., Cole, T. J., Bruner, D., & Treleaven, P. (2008). Body shape in American and British adults: between-country and inter-ethnic comparisons. *International Journal of Obesity, 32,* 152–159.

Wertheimer, A. O. (1994). The first amendment distinction between conduct and content: A conceptual framework for understanding fighting words jurisprudence. *Fordham Law Review, 63,* 793–851.

White, R. (2012). Police Cooperation'. In M. E. Beare (Ed.), *Encyclopedia of Transnational Crime and Justice.* Thousand Oaks: Sage.

Winkler, A. (2013). *Gunfight.* New York: W. W. Norton.

Wolf, P., Troxler, P., Kocher, P. Y., Harboe, J., & Gaudenz, U. (2014). Sharing is sparing: Open knowledge sharing in Fab Labs. *Journal of Peer Production, (5)* Shared Machine Shops.

Wong, J. (2011). *Penrose triangle illusion.* MakerBot Thingiverse. http://www.thingiverse.com/thing:6474. Accessed 11 Sept 2015.

Worthington, E. (2014). 3D printed guns: PUP introduces Queensland bill to regulate digitally generated firearms. *ABC*. http://www.abc.net.au/news/2014-05-23/3-d-printed-guns-palmer-party-introduces-qld-bill-3d-firearms/5472566. Accessed 12 Sept 2015.

Wright, A. and De Filippi, P. (2015). Decentralized blockchain technology and the rise of lex cryptographia. *SSRN Working Paper*. http://ssrn.com/abstract=2580664. Accessed 3 November 2015.

Zapparoni, R. (2004). Propertising identity: Understanding the United States rights of publicity and its implications – Some lessons for Australia. *Melbourne University Law Review, 28,* 690–723.

Zingales, N. (2012). *Digital copyright, "fair access" and the problem of DRM misuse.* Boston College Intellectual Property & Technology Forum.

Zittrain, J. (2008). *The future of the internet and how to stop it.* New Haven: Yale University Press.

INDEX

© The Editor(s) (if applicable) and The Author(s) 2016   117
A. Daly, *Socio-Legal Aspects of the 3D Printing Revolution*,
DOI 10.1057/978-1-137-51556-8